WATER RESOURCES MANAGEMENT SERIES : 2

International Waters of the
Middle East
From Euphrates-Tigris to Nile

To

Dr. Ariel Dinar

with best regards.

Asit K. Biswas

27. 9. 94

Water Resources Management Series

ᴕOURCES MANAGEMENT SERIES : 2

ternational Waters of the Middle East

From Euphrates-Tigris to Nile

Edited by
ASIT K. BISWAS

Sponsored by
United Nations University
International Water Resources Association
With the support of
Sasakawa Peace Foundation
United Nations Environment Programme

OXFORD UNIVERSITY PRESS
BOMBAY DELHI CALCUTTA MADRAS
1994

Oxford University Press, Walton Street, Oxford

Oxford New York Toronto
Delhi Bombay Calcutta Madras Karachi
Kuala Lumpur Singapore Hong Kong Tokyo
Nairobi Dar es Salaam Cape Town
Melbourne Auckland Madrid
and Associates in
Berlin Ibadan

First published 1994
© United Nations University,
International Water Resources Association,
Sasakawa Peace Foundation and
United Nations Environment Programme, 1994

ISBN 0 19 563557 4

Printed in India
Typeset by Alliance Phototypesetters, Pondicherry 605 013
Printed at Rajkamal Electric Press, Delhi 110 033
Published by Neil O'Brien, Oxford University Press
YMCA Library Building, Jai Singh Road, New Delhi 110 001

This book is dedicated to
DR TAKASHI SHIRASU

as a mark of esteem for his contributions to peace
and a token of true regard for a friend

Contents

Contributors

Özden Bilen
Water Resources Engineer and
  Deputy Director General
General Directorate of
  State Hydraulic Works (DSI)
Ankara
Turkey

Asit K. Biswas
Chairman
Middle East Water Commission
International Water Resources
  Association
76 Woodstock Close
Oxford OX2 8DD
England

John Kolars
Professor Emeritus
Department of Near Eastern
  Studies
University of Michigan

Permanent Address:
424 Mission Road
Santa Fe
NM 87501
U.S.A.

Yahia Abdel Mageed
Former Secretary General
U. N. Water Conference
Associated Consultants
P. O. Box 2960
Khartoum
The Sudan

Masuhiro Murakami
Nippon Koei Co. Ltd.
Consulting Engineers
4 Kojimachi
5-chome Chiyoda-ku
Tokyo 102
Japan

Katsumi Musiake
Professor
Institute of Industrial Science
University of Tokyo
7-22-1, Roppongi
Minato-ku
Tokyo 106
Japan

Mostafa Kamal Tolba
Executive Director (1976–1982)
United Nations Environment
  Programme
46 Mosadek Street
Dokki
Cairo
Egypt

Aaron T. Wolf
Assistant Professor of
  Geography
University of Alabama
202 Farrah Hall
Box 870322
Tuscaloosa
AL 35487
U. S. A.

Series Preface:
Water Resources Management

In recent years there has been increasing realization of the importance of water in the continuing well-being and development of mankind. In nearly all countries of the world, ranging from Algeria to Zimbabwe, more and more planners and decision-makers have started to realize the critical importance of efficient water resources management for their sustainable development. Even in an advanced industrialized country like the United States, water availability and its rational management has become a major socio-political issue, especially in the western and south-western parts of the county.

On the basis of extensive analyses carried out, it is now evident that compared to earlier generations of water development projects, new sources of water are becoming scarce, more expensive to develop, require more expertise and technological knowhow for planning, design, implementation and operation, and are contributing to more social and environmental disruptions. Accordingly, it is being increasingly realized that water can no longer be considered to be a cheap resource, which can be profligately used, abused or squandered without noticeable consequences for the future of mankind. Like oil some two decades ago, the day when water could be considered a cheap and plentiful resource is now over in most countries of the world.

If the current trends continue, the situation is likely to deteriorate even further in the future for two important reasons. First, the global population is increasing rapidly, and is likely to continue to do so till about the year 2050, or even beyond. This means more and more water would be required for domestic and industrial uses, agricultural production and hydropower generation for this expanding population. Second, as more and more people attain a higher standard of living, per capita water demand would continue to increase as well. Current analyses indicate that the total global water consumption during the period 1900–2000 is likely to increase ten-fold, and this trend is likely to extend well into the twenty-first century.

In addition, as human population and activities increase, more and more waste products are contaminating available sources of surface water as well as groundwater. Among the major contaminants are untreated or partially treated sewage, agricultural chemicals and industrial effluents. These contaminants are seriously affecting the quality of available sources of water for various uses. Thus, water quality management is becoming increasingly an important concern all over the world.

Another major factor which could affect water management in the future is likely to be increasing delays in implementing new projects. Higher project costs and lack of investment funds would be two major reasons for this delay. Equally, social and environmental reasons would significantly delay project initiation time, certainly more than what has been witnessed in the recent decades.

Since on a long-term basis the amount of water available to any country is limited, the traditional response of the past to increase water availability to meet higher and higher water demands would no longer be a feasible solution in the future. This means that water professionals will come under increasing strain to make the management process more efficient than it has ever been at any time in human history. However, the transition period available to us to significantly improve the water planning and management processes is likely to be short, certainly not more than a decade or at most two. While technological problems, though complex, may prove comparatively easy to solve, economic, political, social and environmental constraints are likely to be more difficult to resolve. Thus a proper approach to the solution of water problems is one of the most difficult challenges facing water management in the twenty-first century.

The Water Resources Management series of books, monographs and state-of-the-art reviews consist of authoritative texts written by some of the world's leading experts in their field. The series as a whole will consider all aspects of water—quantity and quality, surface water and groundwater—from the viewpoints of all the major associated disciplines, i.e. technical, economic, social, environmental, legal, health and political. It will also consider all types of water use: domestic, industrial, agricultural, hydropower generation, navigation, recreation and wildlife enhancement. Individual books may of course have more specific focus. The books of the series would not only be of direct interest to students and professors but also to all professionals associated with water resources planning and management.

ASIT K. BISWAS
International Development Centre
University of Oxford
76 Woodstock Close
Oxford, England

Preface

More than two thousand years ago, the eminent Greek philosopher Pindar said water is the best of all things. For the countries of the Middle East, which are currently facing serious water scarcity conditions, Pindar's view under the existing conditions can probably be best viewed as an understatement.

Because of the critical importance of water for the further socio-economic development of the Middle East, and the complexity of the water management process of the various countries involved, the Committee on International Waters of the International Water Resources Association (IWRA) convened a Middle East Water Forum in Cairo, 7–9 February 1993. As the Chairman of the Committee, I had the privilege to convene this important Forum.

Participation to the Forum was strictly restricted to well-known experts by invitation only. Twenty-seven leading authorities on Middle East waters were carefully selected and invited because of their acknowledged expertise and interest in this field. Participants to the Forum were invited in their private capacities for a free and frank exchange of ideas, opinions and facts.

The Forum was officially opened by Dr Mostafa K. Tolba, former Executive Director of the UNEP. In his opening address he pointed out that before the current changes in Eastern Europe and the former USSR, the number of major river and lake basins shared by two or more countries was estimated to be 214. There are a number of important international river basins in the Middle East region: Euphrates-Tigris, Jordan, Yarmouk and the Nile. In addition, there are many underground aquifers which are shared by two or more countries.

Past and recent experiences indicate that the complex and politically sensitive issues of international river basins cannot be resolved by individual countries unilaterally. They would require genuine cooperation between countries as well as understanding and appreciation of each other's needs. Such cooperation could manifest itself in the form of joint action plans, joint commissions or even treaties. What is needed is regional or sub-regional cooperation. The regions in this context are not necessarily political regions but rather ecological regions like a lake or river basin. What is now required is not cooperation in dividing amounts of water but genuine cooperation in the implementation of agreed plans for the integrated development of the whole basin for the benefit of each country of that region. Such cooperation can come only through availability of reliable facts and consideration of all feasible options. The complex issues can often be best considered and discussed in a closed non-governmental professional forum like the present one where leading technocrats of the region and renowned experts on Middle East Waters from outside the region were specially invited to discuss these multi-faceted issues. Dr Tolba also pointed out that the region had high expectations from the deliberations of such a high level of experts.

The Forum agreed to spend about half the time available for discussion of the background circulated and the second half on brainstorming on the issues raised and other relevant facts which warrant consideration.

The most exceptional aspect of the Forum was the successful exercise in international relations it proved to be. The participants discussed in a scientific and objective manner the complex problems of sharing limited water resources available in a very arid region, with a population base rapidly expanding due to both natural causes and migration. There were many areas where there were sharp differences of opinion, but these were outlined and discussed in a constructive fashion, without rancour.

The Forum participants primarily focused their attention on facts and figures available, and the various implications of such facts and figures. It is clear that there are many data, whose reliabilities are being questioned by one or more parties. Lack of standardization of the data set has created further misunderstanding. For example, 'dunum', a unit of area, has different values in different Middle East countries.

There was considerable discussion on how mind-sets of decision-makers could be changed by indicating that water need not be the basis of a zero-sum game for the countries concerned in the region. Many participants felt that the present perception needs to be changed since water availability could be increased through different technical options. Equally water use and demand patterns could be dramatically changed through better water conservation techniques, changing cropping patterns, shifting water use from agricultural to other purposes and consideration of other similar alternatives.

Several participants felt that many of the implicit but fundamental assumptions on water availability and demand patterns need to be seriously re-examined, not only in the context of the Middle East but also for other arid and semi-arid regions of the world.

Political issues and considerations were mentioned, where these were necessary. It is clear that water policies in a very arid region with limited water supply cannot be divorced or discussed separately from the political issues facing the sovereign states. The overall atmosphere during the entire Forum was generally amiable, even when the participants disagreed with each other.

Since 17 of the 27 participants in the Forum are now associated with the bilateral and the multilateral negotiations for the Middle East Peace Talks, the Forum unanimously agreed that it should not make specific recommendations which could affect the progress of the Peace Talks. However, the simple fact that these negotiators could meet each other in an informal setting and get a better understanding of the reasons or rationale behind many of the negotiating situations, can only augur well for the future discussions on water within the context of the Middle East Peace Talks.

The Forum participants had not had an opportunity to interact with each other in a personal and informal manner before. The unanimous conclusion of the participants was that they found the Forum most useful in terms of developing personal contacts, new ideas and better understanding of many technical facts. They felt professional NGOs like IWRA have a major role to play in the management of international waters. They encouraged IWRA to convene a similar well-focused and well-planned Forum for South and South-East Asia and later for Africa. Such Forums would facilitate negotiations and development of international water bodies in the regions concerned.

Preparation and convening of a well-focused Forum on a very complex and difficult subject is not an easy task under the best of circumstances. The present Forum was no exception. While IWRA was considering convening of the Forum, I had an opportunity to discuss this possibility with Dr Roland Fuchs, Vice-Rector of the United Nations University, Tokyo. He enthusiastically endorsed the idea, and encouraged me to proceed with the organization of the Forum as soon as possible. Both Dr Fuchs and Dr Juha Uitto of the U.N. University (UNU) have consistently supported me with the convening of the Forum and the resulting follow-up activities.

I then had the opportunity to discuss the Forum, almost simultaneously, with Dr Mostafa Kamal Tolba and Dr Habib N. El-Habr of the United Nations Environment Programme (UNEP) and Dr Takashi Shirasu of the Sasakawa Peace Foundation (SPF) during my frequent visits to Nairobi and Tokyo. All of them strongly supported the idea. The main question I faced from our three sponsors—United Nations University, United Nations Environment Programme and the Sasakawa Peace Foundation

—was how soon we could realistically convene a well-organized Forum with the right participants which could produce concrete results, and not on its need. I am really grateful for all the support and encouragement I have received, and continue to receive, from Drs Fuchs, Uitto, Tolba, El-Habr and Shirasu on the Forum-related activities.

Two other individuals helped me extensively in the organization of the Forum, my wife Margaret and Dr Jerome Delli Priscoli of the Institute of Water Resources, Fort Belvoir, U.S.A. Dr Priscoli also acted as a superb facilitator at the Forum. Their continuous advice further contributed to the remarkable success of the Forum.

Last but not least I am most grateful to all the participants to the Forum. I do not recall a single instance during the past three decades when all the high-level experts formally invited to a meeting have promptly accepted the invitation. This is also the first time in the history of IWRA, when *all* members of *any* Committee participated in an event. Such enthusiasm clearly indicated the need for and importance of the Forum.

The background papers specifically commissioned for the Forum and the opening address of Dr M. K. Tolba form the basis of this book. It should be noted that the authors were invited to prepare their papers in their personal capacities, and thus the opinions expressed are those of the authors, and not necessarily those of the sponsors.

Through the courtesy of our three sponsors—UNU, UNEP and SPF—the book is being made widely available to all the major players of the Middle East water issues, both inside and outside the region. I have no doubt that this will contribute to better understanding of the various Middle East water issues nationally, regionally and globally. It should further facilitate improved dialogue between the various players.

What next? Because of the remarkable success of the Forum, and through the kind support of the Sasakawa Peace Foundation, we have now constituted a blue-ribbon Middle East Water Commission, with myself as the Chairman, and Prof. John Kolars, Dr Masahiro Murakami, Prof. John Waterbury and Prof. Aaron T. Wolf as members. The first meeting of the Commission was held in Santa Fe, U.S.A., 28–30 September 1993. The Commission is studying the various dimensions and aspects of the Middle East water issues. The Commission report is expected to be finalized in late 1994. It is planned to make this report widely available as an important and useful contribution to sustainable water management of the region. Since no lasting peace in the Middle East is possible without a just and rational agreement on its water resources, we hope that the results of the Middle East Water Forum and the work of this Commission will play an important and perceptible role in bringing peace and prosperity to the region.

Asit K. Biswas, Chairman
Middle East Water Commission
International Water Resources Association
76 Woodstock Close
Oxford, England

1 / Middle East Water Issues: Action and Political Will

MOSTAFA KAMAL TOLBA

For the past ten years, and in particular over the past five years, I have been persistently calling for action rather than talk. I sincerely hope that this meeting is going to set action in motion. We have plenty of recommendations from global, regional and other international conferences, meetings, seminars and the like. We have plans of action at the global, regional and national levels. We know the facts and what needs to be done. But we have yet to see real action.

We all know the following facts about water in the Middle East:

1. Fresh water is a key factor in development, particularly in arid and semi-arid countries which include countries of the Middle East. In this region, water security, like food security, is a matter of survival.

2. The availability of fresh water is very unevenly distributed in the Middle East. The relative availability of internal renewable water resources ranges from an extremely low level of almost zero m³ per capita to a high level of almost 7000 m³. More disturbing are the uncertainties surrounding the potential impacts of climatic change and global warming on the rain patterns and hence on water resources.

3. Against all this, total water use has increased nearly tenfold over the last 100 years. New sources of water have become more scarce and more expensive to develop, while competition amongst the various users has increased. This situation will worsen over the next 20–30 years.

4. The severity of water shortage is further aggravated by a steady deterioration in water quality. Pollution from industry, urban wastewater and agricultural run-off reduces the fitness of fresh water sources in this region.

5. Where groundwater is used as the source of irrigation water, a steady decline in groundwater levels has often followed.

6. Inefficient irrigation systems are responsible for the majority of water losses. 40–60 per cent of water abstracted from rivers often does not reach the agricultural fields. Instead it is lost, mainly through seepage. Inappropriate irrigation practices result in the development of salinity and water-logging. And irrigation and overgrazing by livestock significantly increases soil erosion. Much of the soil eroded contributes to higher sediment loads in watercourses. This in turn can result in higher

than expected sedimentation rates for reservoirs, with concomitant serious economic losses.

7. When industrial wastewater, which is very heterogeneous and uneven in composition, is discharged into surface water, or mixed with municipal wastewater for irrigation, it creates serious environmental problems.

8. There is a whole range of diseases associated with water which are prevalent in the region. These include waterborne (bacterial, viral, parasitic), water-washed (e.g. enteric, skin) and other water-related (e.g. malaria, schistosomiasis) diseases.

There are known tested solutions to all these problems. Agenda 21 which was adopted at the Earth Summit in June 1992 dealt in detail with what needs to be done in the area of fresh water resources in different parts of the world. There are recipes for preventing the worst from happening. I know that these recipes are much easier said than done. But they *can* be applied. What is necessary now is for each government to set specific achievable targets and commit itself to achieving these targets through specific actions over specific periods of time, identifying the actors, committing national financial and human resources, defining the needed external help and identifying the potential donors for such help.

There is no blueprint for this. The challenge for us in this meeting is to design options which governments in the Middle East can choose from or adapt. However, to do this, most countries in the region need to remove the major constraints on efficient water management: the weaknesses of the institutions concerned. As a general rule, it can be said that most water institutions in the region—and they are often amongst the first institutions to be established in these countries—need significant strengthening in order to cope with emerging water management problems brought about by new agricultural crops, large-scale water management projects, and changed social, cultural and political environments. In addition, in order that water can be managed in its totality in a rational fashion, inter-institutional collaboration has to be substantially improved. Currently, in nearly all the countries of the region, water-related policies have been developed in a fragmented fashion by a host of institutions. For example, irrigation is handled by ministries of irrigation or water resources, water supply by municipalities, hydroelectric power by ministries of energy, navigation by ministries of transport, environment by ministries of environment, and health by ministries of health. Lack of coordination, and often intense rivalry between ministries, have meant that water policies have generally been suboptimal. Without institutional rationalization and strengthening, optimal water management is simply not possible in the future.

In addition, this region has two delicate problems to face:

1. Water pricing and cost recovery: Adam Smith once pointed out that there appeared to be a paradox in the fact that fresh water, which is vital for the sustenance of all life, costs nothing, whereas diamonds, which are vital for nothing at all, cost a lot. In other words, so long as there is an unlimited supply of fresh water, there is very little reason for water to affect the economic or legal workings of civilized society. Unfortunately, the theory of unlimited supply is no longer valid in the case of water. As a result, since the 1980s, the issue of water pricing and cost recovery has been a major discussion topic in many national and international fora. The view was put forward that if economically realistic water prices could be charged, farmers and other consumers would become rational optimizers, which would contribute substantially to efficient water use. Furthermore, if government departments could receive the extra revenue generated by water pricing, they could operate and maintain their water systems much more efficiently.

By the early 1990s, it was increasingly realized that one fundamental issue has to be considered before water pricing becomes an attractive policy instrument. Water pricing has thus far been viewed primarily as an economic instrument: its socio-political implications in developing countries have generally not been understood. However, water has traditionally been subsidized to achieve very specific socio-political objectives of food security, and increasing the health standards and incomes of the rural poor, especially women. So, if economic water pricing is to be introduced, other policy instruments must be developed to achieve the socio-political objectives.

It is thus important to recognize that water pricing can only be successfully applied after thorough and frank discussions involving all those concerned: governments, parliaments, the scientific community, economists and the various groups of consumers.

2. The second and probably more sensitive or explosive issue is that of shared water resources.

Before the changes in Eastern Europe and the former USSR, the number of rivers and lake basins shared by two or more countries was 214. A number of very important ones among these are in the Middle East: the Rivers Nile, Euphrates, Yarmouk, Litani, Jordan, and others. In addition, there are shared underground water aquifers like the Nubian Sandstone aquifer and the Arabian Peninsula aquifer. The development and management of these shared water resources pose special challenges, which sometimes become explosive political issues. As the demand for water and hydroelectric power increases, and exclusively national sources of water are fully developed, the only major new sources are likely to be international.

Past and current experience points to the fact that these complex and politically sensitive issues are not resolved unilaterally. They require genuine cooperation based on a genuine understanding of other peoples' needs. Such cooperation manifests itself in the form of joint action plans, joint commissions or even treaties. Based on diagnostic studies, a number of action plans have been formulated in various parts of the world, which have the concurrence of all countries involved. One such plan is being prepared for the Nile. In my view this is the only way to tackle these political time-bombs. Talking about dividing the amount of water will not help. Tremendous deforestation of watersheds compounded by the expected climatic change with its impact on rain patterns will seriously affect the availability of water itself. What is needed is regional or subregional cooperation. I am not talking about political regions, I am talking of ecological regions, a river or a lake basin. And I am not talking about cooperation in dividing the water, but cooperation in the implementation of agreed plans for the integrated development of whole basins for the benefit of everybody.

In 1982, on the occasion of the 10th Anniversary of the Stockholm Conference on the Human Environment, I had said, 'Water will take its place next to energy as a major political issue in the next decade.' I maintain what I said then, and I claim that the Middle East is one of the most sensitive areas in this respect. A breakthrough in this field can probably play a major role in securing the success of the peace process in the region. We have been so used to glibly using the words 'Political Will'. I do not think political will comes out of a vacuum. It comes out of the relevant and reliable facts provided by scientists or technicians, and the options for possible solutions designed by them.

You are all technicians of the highest level, and I understand that a large number of you are involved, one way or another, in the Middle East Peace talks on the issue of water. I believe that the collective wisdom of this outstanding gathering is capable of coming up with what is specifically needed: reliable facts and feasible options. We expect no less from you.

I thank you and I now declare open the Middle East Water Forum.

2 / A Hydropolitical History of the Nile, Jordan and Euphrates River Basins

AARON T. WOLF

In 1876, John Wesley Powell, the leader of the first organized expedition down the Colorado River, submitted his *Report on the Lands of the Arid Region of the United States* to Congress. Among his observations on U.S. settlement policies in the deserts of the Southwest, was his belief, as described by Marc Reisner (1986), that

> state boundaries were often nonsensical . . . In the West, where the one thing that really mattered was water, states should logically be formed around watersheds . . . To divide the West any other way was to sow the future with rivalries, jealousies, and bitter squabbles whose fruits would contribute solely to the nourishment of lawyers.

The same might belatedly be said about the national boundaries of the Middle East. The difference, of course, is that in this region, conflicts between states have deep historical roots, and are more often settled on the battlefield than in the courtroom.

The basins of the Nile, Jordan and Euphrates Rivers, with all their competing national and economic pressures, provide clear examples of the strategic importance of water as a scarce resource. What follows is a brief summary of the history of water conflict and cooperation between the riparians of these river basins, with an eye towards how the lessons of that history may provide guidelines for the future.

In 1993, we stand at a crucial fork in the stream of hydropolitical events in the Middle East. As regional peace negotiations move forward incrementally, only to be lurched backwards by the inexorable forces of regional politics, we have two choices for planning and development of the region's river systems. We can move along the turbulent branch of increased international competition, resulting in rationing, degraded water quality, and irreversible damage to the fragile groundwater systems and surface reservoirs, or we can use the lessons of more than 100 years of water-induced tensions and attempts at collaboration to help guide us along the more pacific, albeit somewhat less familiar, branch towards cooperation, basin-wide management, and the most efficient use of every drop of this precious and scarce natural resource.

The arguments in favour of the likelihood of increased competition are well-known and exemplified in the statistics for each basin in the accompanying papers in this volume. In general, the converging forces which could precipitate water conflict in the Middle East can be categorized as follows (after U.S. Army Corps of Engineers 1991, pp. 3–7):

- flow variation in time and space;
- population growth;
- inefficient agricultural practices;
- variable precipitation;
- increasing water use with higher standard of living;
- decreasing groundwater availability;
- degrading water quality;
- environmental impacts of water use;
- increasing interdependency;
- weak institutional frameworks for water management; and
- the threat of global warming to water availability.

The disheartening result of these converging forces is that there presently is either an ongoing or an impending conflict over water quantity and/or quality on practically every border between co-riparians along each of the three river-ways described in this paper.

Yet, as Naff and Matson (1984, p. 3) point out, some of the same properties which make water a potential flashpoint for international conflict also allow for the possibility of regional cooperation:

Paradoxically, these very complexities and the virulent danger of hostilities engendered by hydrological problems have often tended to compel cooperation where other non-water antagonisms have degenerated into warfare. Thus, water as an impulsion toward conflict carries its own corollary, being as well an impetus toward cooperation.

This corollary leaves room for optimism. Since watershed planning lends itself to a regional approach, and since issues of water are inextricably linked to vital national issues of security and immigration, resolving conflicts over water may become the most tractable of the issues to be dealt with, providing the opportunity for the confidence-building steps necessary to reach accord over other, more contentious, issues as well. As we will see, people who will not talk about history or politics do, when their lives and economies depend on it, talk about water.

The choice of which direction we take—that of hydro-conflict or that of hydro-cooperation—is a pressing one. Frey (1992, p. 5) cites the Catastrophe Theory, which describes how small changes in a social structure, once begun, can develop and increase quickly, much like the effects of resonating sound waves amplifying to shatter a wine-glass, to depict the cybernetic repercussions of delaying the choice between conflict and cooperation:

The tension and threat (of transnational water shortage) can apparently be resolved either by sharply escalating the conflict or by accepting the necessity of some form of cooperation. Dire conditions promote cooperation, but those same conditions also make severe conflict more likely.

Competition, then, begets ill will which increases competition while, conversely, cooperation encourages better relations which create an environment conducive to increased cooperation. In the worst case, as Meir Ben-Meir, a former Israeli Water Commissioner, has put it, 'If the people in the region are not clever enough to discuss a mutual solution to the problem of water scarcity, then war is unavoidable' (cited in *London Times*, 21 February 1989).

As we weigh how we might tip the cybernetic scale to the side of increasing cooperation, it is worth investigating the history of water-related conflict and cooperation. Good planning for the future, after all, is often founded on a firm understanding of the past. The following section examines the hydropolitical history of the Nile, the Jordan, and the Euphrates basins, with just such an eye towards the future.

A caveat: because of space limitations, what follows is meant as a summary of hydropolitical events only, with some important events excluded. The reader interested in more detail is referred to Naff and Matson (1984) for an excellent overview of the three river basins; to Waterbury (1979) for the hydropolitics of the Nile; to Wolf (1992) for a more thorough history of the Jordan; and to Kolars and Mitchell (1991) for more detail on the Euphrates.

HISTORY—WATER CONFLICT AND COOPERATION

From the origins of civilization in the Middle East, the limits and fluctuations of water resources have played a role in shaping political forces and national boundaries. Water availability helped determine both where and how people lived, and influenced the way in which they related to each other. Issues of water conflict and cooperation have become especially intense with the growing nationalist feelings and populations of this century. These issues are also relevant to current conflict—particularly between Israel, Jordan and the Palestinians on the

West Bank and Gaza—but they may offer new opportunities for dialogue as well.

As the relationship between the water resources and political events
in the region is described below, it should be kept firmly in mind that
nothing described here happened in a political vacuum. Of all the
myriad of geo-political and strategic forces surrounding each of these
developments, only those relating water resources to conflict or
cooperation are extracted here for examination.

A. The Emergence of Agriculture and Nationalism

Living as they do in a transition zone between Mediterranean subtropical
and arid climates, the people in and around the major watersheds of the
Middle East have always been aware of the limits imposed by scarce
water resources. Settlements sprang up in fertile valleys or near large,
permanent wells, and trade routes were established from oasis to oasis.
In ancient times, cycles of weather patterns had occasionally profound
effects on the course of history. Recent research suggests that climatic
changes 10,000 years ago, which caused the average weather patterns
around the Dead Sea to become warmer and drier, may have been an
important factor in the birth of agriculture (Hole and McCorriston, as
reported in the *New York Times*, 2 April 1991). The Natufians of the
Jordan Valley, it has been suggested, found that by planting wild
cereals they could overcome the increasing summertime food shortages
of a drying climate.

It is also becoming increasingly accepted that a similar climatic
drying around 4000 years ago was responsible for the movement of
groups of pastoralists from the marginal lands of the Syrian and
Jordanian steppes, as well as the Negev and Sinai Deserts, because the
marginal land no longer provided enough feed for their herds, into the
more fertile coastal areas of the Eastern Mediterranean. Together, as
these groups shifted from sheep herding to agriculture, they coalesced
into a political/religious entity later to become known as the Israelites.[1]

Even in biblical times, variations in water supply had their impact on
the region's history. It was drought, for example, that drove Jacob and his
family to Egypt, an event which led to years of slavery and, finally, to the
consolidation of the Israelite tribes 400 years later (Genesis 41). And,
even then, the waters of the Jordan were occasionally intertwined with
military strategy as, for instance, when Joshua directed his priests to stem

[1] Summarized in lectures at the University of Wisconsin, Madison, by Anson Rainey, 2
February 1992; and by Lawrence Sinclair, 18 March 1992.

the river's flow with the power of the Ark of the Covenant while he and his army marched across the dry riverbed to attack Jericho (Joshua 4).

Further east, the fluctuations of the Euphrates gave rise to legend (the flood experienced by Noah is thought to have centred its devastation around the Babylonian city of Ur, submerging the southern part of the Euphrates for about 150 days), to extensive water law (the code of King Hammurabi contains as many as 300 sections dealing with irrigation), and to extensive irrigation and flood control works (the dam at Shalalat, on a tributary of the Tigris, built in the 7th century BCE, still stands today (El-Yussif 1983). The irrigation works of ancient Egypt are, of course, no less famous—it is thought that the art of field surveying was invented for their development.

National changes are not restricted to a drying climate. In an exhaustive study of the relationship between the ancient peoples of the Mid-east and their water, Issar (1990) suggests that favourable climatic conditions, with rainfall in the Negev 50 per cent greater than today's, may have contributed to the success of several national entities in the region from about 200 BCE to 400 CE.[2] This was a period in which the Roman Empire included much of the Mid-east, the monastic Dead Sea sect (possibly the Essenes) thrived around the area of Qumran and, further south, the Nabateans extended their hold over the spice trade routes from Arabia to the ports along the Mediterranean Coast. The Nabateans, with cities across the Negev Desert and a stunning capital at Petra, were particularly adept at intensively managing each drop from the rare rain events of their arid territory (Issar 1990, pp. 178–81).

Issar concludes his study with the intriguing speculation that, once the climate again began to dry in the fifth to seventh centuries CE, the inhabitants of the desiccating Arabian Peninsula may have found incentive to search for a more hospitable environment, resulting in the Muslim expansion across the Mid-east, North Africa, and into Spain:

Was this burning religious zeal of the Moslems made fiercer by the droughts which struck the northern and central parts of their peninsula? Did this drying up also weaken the countries of the Fertile Crescent guarding what was left of the Roman Empire . . . ? (Issar 1990, p. 188.)

In the centuries since, the inhabitants of the region and the conquering nations which came and went have lived mostly within the limits of their water resources, using combinations of surface water and well water for survival and livelihood (Beaumont 1991, p. 1). But just as changing

---

[2] BCE, Before the Common Era; CE, Common Era.

amounts of water availability in the Mid-east may have contributed to the formation of both the Jewish and Arab nations millennia ago, conflicting interpretations of how to overcome those limits have also been a factor in competition and conflict as their respective nationalisms began to re-emerge on the same soil in this century.

When, after the first Zionist Congress in Basle in 1897, the idea of creating a Jewish State in Palestine, which by then had been under Ottoman rule for 400 years, began to crystallize in the plans of European Jewry, Theodore Herzl, considered the father of modern Zionism, travelled to the region to see what practical possibilities existed. In Jerusalem, Herzl met with the German Kaiser, whose influence with the Ottoman Sultan he sought to enlist. Barbara Tuchman describes the 1896 meeting outside the Mikveh Israel colony:

The Kaiser rode up, guarded by Turkish outriders, reigned in his horse, shook hands with Herzl to the awe of the crowd, remarked on the heat, pronounced Palestine a land with a future, 'but it needs water, plenty of water', shook hands again, and rode off (Tuchman 1956, p. 291).

Frustrated by the lack of enthusiasm for Jewish settlement on the part of the Turks, Herzl turned to the British, whose control of Egypt extended into the northern Sinai peninsula. In 1902, Herzl suggested to Joseph Chamberlain, the British colonial secretary, that Jewish colonization and massive irrigation of the territory around El-Arish, in the northern Sinai peninsula, would create a 'buffer state' between Egypt and Turkey, helping to protect British interests in the Suez Canal (Ra'anan 1955, pp. 36–7). Although Chamberlain was supportive, Lord Cromer, head of the Anglo-Egyptian Administration in Cairo, was sceptical of the chances for success of Jewish colonization and wary of intimidating the Turks, with whom the legal boundaries in the area were unclear. Cromer finally vetoed the project in 1903, claiming that Nile water, which would be necessary for irrigation, could not be spared.

The British guaranteed Egyptian interests from up-river as well, which, combined with a lack of pressure on the water resources, resulted in a general absence of hydropolitical conflict along the Nile until the early 1900s. This stability began to be strained with a relative shortage of cotton on the world market around the turn of the century, which put pressure on Egypt and the Sudan, then under a British-Egyptian condominium, to turn to this summer crop, requiring peren-nial irrigation rather than the traditional flood-fed methods. The need for summer water and flood control led to an intensive period of water development along the Nile, with proponents of Egyptian and Sudanese interests occasionally clashing within the British foreign office over whether the emphasis for development ought to be further up-stream or down (Naff and Matson 1984, pp. 141–2).

The most extensive scheme for comprehensive water development along the Nile, proposed in 1920 and now known as the Century Storage Scheme, included a storage facility on the Uganda-Sudan border, a dam at Sennar to irrigate the Gezira region south of Khartoum, and another dam on the White Nile to hold summer flood water for Egypt. The plan worried some Egyptians, and was criticized by nationalists, because all the major control structures would be beyond Egyptian territory and authority. A Nile Project Commission, including a representative of the Indian government as chair, a nominee from Cambridge University, and a nominee from the U.S. government, mediated the conflict in 1920, and, though they estimated Egyptian needs at 58,000 MCM/year, with the rest of the annual flow to be allocated to the Sudan, no agreement was reached (Krishna in Starr and Stoll 1988, p. 25; Naff and Matson 1984, p. 143).

Meanwhile, in Palestine, even without commitments for independent nations, both Jewish and Arab populations were beginning to swell by the turn of the century, the former in waves of immigration from Yemen as well as from Europe, and the latter attracted to new regional prosperity from other parts of the Arab world (Sachar 1969; McCarthy 1990). According to Justin McCarthy (1990), Palestine had 340,000 people in 1878 and 722,000 by 1915.

During World War I, as it became clear that the Ottoman Empire was crumbling, the heirs-apparent began to jockey for positions of favour with the inhabitants of the region. The French had made inroads with the Maronite Catholics of Lebanon and therefore focused on the northern territories of Lebanon and Syria. The British, meanwhile, began to seek coalition with the Arabs from Palestine and Arabia—whose military assistance against the Turks they desired—and with the Jews of Palestine, both for military assistance and for the political support of diaspora Jewry (Ra'anan 1955).

As the course of the war became clear, French and British, Arabs and Jews, all began to refine their territorial interests. And the location of the region's scarce water resources was a critical factor in the decision-making process of each party, particularly in Palestine. (See Figure 1— Border Proposals, 1919–1947.)

On 9 March 1916, the Sykes-Picot Agreement was clandestinely

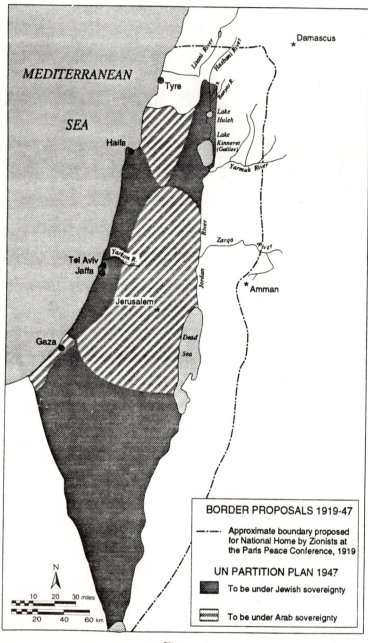

Figure 1

signed between the British and French, dividing the Mid-east into regions which would be designated as French (including Lebanon and the northern Galilee), French-influence (Syria), British (Egypt, Iraq and the port of Haifa/Acre), British-influence (northern Saudi Arabia and Jordan), and international (the remainder of Palestine), (Ra'anan 1955, p. 68).

The spheres of influence of the Sykes-Picot Agreement would have left the watersheds in the region divided in a particularly convoluted manner: the Litani and the Jordan headwaters to just south of the Huleh region would be French; the Sea of Galilee would be divided between international and French zones; the Yarmouk valley would be split between British and French; and the lower stem of the Jordan would be international on the west bank and British on the east. The Euphrates would have fared little better, being the divide between the zones of French and British influence, and the British zone around and south of Baghdad.

Because of these divisions, and because there is no mention of water per se in the literature on these negotiations, we suggest that other factors such as the locations of rail and oil lines, the presence of holy places, and political debts and alliances took precedence, and water was not an issue up to this point in the border demarcation process (see Ra'anan 1955 and Fromkin 1989 for thorough discussions of these other factors). After the Sykes-Picot agreement, however, and as the outcome of the war began to become clear, each entity with national claims in the region increasingly included water resources in its geographic reasoning, particularly after the end of World War I in 1918.

On 2 November 1917, the Balfour Declaration was approved by the British Cabinet:

His Majesty's Government view with favour the establishment in Palestine of a national home for the Jewish people, and will use their best endeavours to facilitate the achievement of this object, it being clearly understood that nothing shall be done which may prejudice the civil and religious rights of existing non-Jewish communities in Palestine, or the rights and political status enjoyed by Jews in any other country (reproduced in Friedman 1987, vol. 8).

Conflicting interpretations of what was meant by 'national home', or even by 'Palestine' (at the time including both sides of the Jordan River), and the apparent contradiction between 'facilitating this object' and 'not prejudicing the . . . rights of existing non-Jewish communities', would lead to contention for years to come.

Because of British conquests in Palestine by the end of 1918, the

British no longer felt overly obligated to the French and new political interests began to be incorporated in the delineation of borders. Although not acceding totally to Zionist requests, the British did deviate from the Sykes-Picot line and adopted the biblical 'Dan to Beersheba' for Palestine, as based on a map of 'Palestine under David and Solomon' (Hof 1985, p. 11), in negotiations with the French over the temporary boundaries of 'Occupied Enemy Territorial Administrations (OETA)', but held open the possibility that,

Whatever the administrative sub-divisions, we must recover for Palestine, be it Hebrew or Arab, the boundaries up to the Litani on the coast, and across to Banias, the old Dan, or Huleh in the interior (Lord Curzon, cited in Ingrams 1972, p. 49).

In 1919, with the war over, and as preparations for the Paris peace talks began at Versailles, border requirements were refined by each side, Zionist and Arab.

1. Zionist Position

The Zionists began to formulate their desired boundaries for the 'national home', to be determined by three criteria: historic, strategic and economic considerations (Zionist publications cited in Ra'anan 1955, p. 86).

Historic concerns coincided roughly with British allusions to the biblical 'Dan to Beersheba'. These were considered minimum requirements which had to be supplemented with territory which would allow military and economic security. Military security required desert areas to the south and east as well as the Beka'a valley, a gateway in the north between the Lebanon Mountains and Mount Hermon.

Economic security was defined by water resources. The entire Zionist programme of immigration and settlement required water for large-scale irrigation and, in a land with no fossil fuels, for hydro-power. The plans were 'completely dependent' on the acquisition of the 'headwaters of the Jordan, the Litani River, the snows of Hermon, the Yarmouk and its tributaries, and the Jabbok' (Ra'anan 1955, p. 87).

The guiding force in refining the thinking on the necessary boundaries was Aaron Aaronsohn. In charge of an agricultural experimental station at Atlit on the Mediterranean coast, Aaronsohn's research focused on weather-resistant crops and dry-farming techniques. Convinced that the modern agricultural practices which would fuel Jewish immigration were incompatible with 'the slothful, brutish Ottoman regime' (Sachar 1979, p. 103), he concluded that Zionist settlement objectives required alliance with the incoming Allied Forces. Aaronsohn initiated contact with the British to establish a Jewish spy network in Palestine, which would report on Turkish positions and troop movements. Perhaps because of his training both in agriculture and in security matters, he became the first to delineate boundary requirements specifically on future water needs. Aaronsohn's 'The Boundaries of Palestine' (27 January 1919, unpublished, Zionist Archives), drafted in less than a day, argued that,

In Palestine, like in any other country of arid and semi-arid character, animal and plant life and, therefore, the whole economic life directly depends on the available water supply. It is, therefore, of vital importance not only to secure all water resources already feeding the country, but also to insure the possession of whatever can conserve and increase these water—and eventually power—resources. The main water resources of Palestine come from the North, from the two mighty mountain-masses—the Lebanon range, and the Hermon . . .

The boundary of Palestine in the North and in the North East is thus dictated by the extension of the Hermon range and its water basins. The only scientific and economic correct lines of delineation are the water-sheds.

Aaronsohn then described the proposed boundaries in detail, as delineated by the local watersheds. He acknowledged that, with the exception of the Litani, the Lebanon range sends no important water source towards Palestine and, 'cannot, therefore, be claimed to be a "Spring of Life" to the country'. It is the Hermon, he argued, that is, 'the real "Father of Waters" and [Palestine] cannot be severed from it without striking at the very root of its economic life'.

Aaronsohn's rationale and boundary proposals were adopted by the official Zionist delegation to the Peace Conference, led by Chaim Weizmann. The 'Boundaries' section of the 'Statement of the Zionist Organization Regarding Palestine', which paraphrased Aaronsohn, read, in part:

The economic life of Palestine, like that of every other semi-arid country depends on the available water supply. It is therefore, of vital importance not only to secure all water resources already feeding the country, but also to be able to conserve and control them at their sources.

The Hermon is Palestine's real 'Father of Waters' and cannot be severed from it without striking at the very root of its economic life . . . Some international arrangement must be made whereby the riparian rights of the people dwelling south of the Litani River may be fully protected. Properly cared for these head waters can be made to serve in the development of the Lebanon as well as of Palestine (Proposals dated 3 February 1919, Weizmann Letters 1983, Appendix II).

Interestingly, Aaronsohn thought his ideas had been badly mangled in the Proposals, perhaps because he was not included in the final drafting. In an angry letter to Weizmann, he complained that the draft was, '*a disgrace and a calamity*' (emphasis Aaronsohn's), and expressed shock that, for one of the delegates, 'a "watershed" is the same as a "thalweg". Incredible, but true' (unpublished letter, 16 February 1919, Weizmann Archives).

In June 1919, Aaronsohn died in a plane crash on his way to the Peace Conference, and the Zionist proposals were submitted without revision. Nevertheless, the importance of the region's water resources remained embedded in the thinking of the Zionist establishment. 'So far as the northern boundary is concerned,' wrote Chaim Weizmann later that year, 'the guiding consideration with us has been economic, and "economic" in this connection means "water supply" ' (Weizmann Letters, 18 September 1919).

2. Arab Position

The Arab delegation to the Peace Conference was led by the Emir Feisal, younger son of Emir Hussein of the Hejaz. Working with T. E. Lawrence, Hussein and his sons had led Arab irregulars against the Turks in Arabia and Eastern Palestine. After the war, Feisal had developed a relationship with Chaim Weizmann as both prepared for the Peace Conference. After a meeting in 1918, Feisal said in an interview,

The two main branches of the Semitic family, Arabs and Jews, understand one another, and I hope that as a result of interchange of ideas at the Peace Conference, which will be guided by ideals of self-determination and nationality, each nation will make definite progress towards the realization of its aspirations (cited in Esco Foundation 1947, p. 139).

Feisal also initially expressed support for Jewish immigration to Palestine, in part because he saw it as useful for his own nationalist aspirations. At a banquet given in his honour by Lord Rothschild in 1918, he pointed out that, 'no state could be built up in the Near East without borrowing from the ideas, knowledge and experience of Europe, and the Jews were the intermediaries who could best translate European experience to suit Arab life' (Esco Foundation 1947, p. 140).

In a meeting later that year, Feisal tried to enlist Weizmann's support against French policies in Syria. Weizmann in turn outlined Zionist aspirations and, 'asserted his respect for Arab communal rights' (Sachar 1969, p. 385). The two also agreed that all water and farm boundary questions should be settled directly between the two parties.

Feisal and Weizmann formalized their understanding to support each other's nationalist ambitions on 3 January 1919, in a document which expressed mutual friendship and recognition of the Balfour Declaration, and stated that,

All necessary measures shall be taken to encourage and stimulate immigration of Jews into Palestine on a large scale, and as quickly as possible to settle Jewish immigrants upon the land through closer settlement and intensive cultivation of the soil. In taking such measures the Arab peasant and tenant farmers shall be protected in their rights, and shall be assisted in forwarding their economic development [original reproduced in Weizmann Letters],

providing, Feisal hand-wrote in the margin, that Arab requests were granted. 'If changes are made,' he wrote, 'I cannot be answerable for failure to carry out this agreement.'

The Arab requests were spelled out in a memorandum dated 1 January 1919. Because the territory in question was so large (including Syria, Mesopotamia and the Arabian Peninsula), geographically diverse and, for the most part, well-watered, it is not surprising that water resources played little role in the Arab deliberations. Based on a combination of level of development and ethnic considerations, Feisal asked that (from Esco Foundation 1947):

– Syria, agriculturally and industrially advanced, and considered politically developed, be allowed to manage her own affairs;

– Mesopotamia, 'underdeveloped and thinly inhabited by semi-nomadic peoples, would have to be buttressed . . . by a great foreign power', but governed by Arabs chosen by the 'selective rather than the elective principle';

– the Hejaz and Arabian Peninsula, mainly tribal areas suited to patriarchal conditions, should retain their complete independence.

Two areas were specifically excluded: Lebanon, 'because the majority of the inhabitants were Christian', and which had its own delegates, and Palestine which because of its 'universal character was left to one side for mutual considerations of all parties interested' (Esco Foundation 1947, p. 138).

Once testimony was heard at Versailles, the decisions were left to the British and the French as the peace talks continued, as to where the boundaries between their mandates would be drawn. At the San Remo Conference in April 1920, agreement was reached where Great Britain was granted the mandates to Palestine and Mesopotamia, and France received the mandate for Syria (including Lebanon). During the remainder of the year, last-minute appeals were made both by the

British and by the Zionists for the inclusion of the Litani in Palestine or, at the least, for the right to divert a portion of the river into the Jordan basin for hydro-power. The French refused, offering a bleak picture of the future without an agreement and suggested, referring to British and Zionist ambiguity about what was meant by a 'National Home', '*Vous barbotterez si vous le voulez, mais vous ne barbotterez pas á nos frais*'[1] (Butler and Bury eds. 1958, cited in Hof 1985).

On 4 December 1920, a final agreement was reached in principle on the boundary issue, which mainly addressed French and British rights to railways and oil pipelines, and incorporated the French proposal for the northern boundaries of six months earlier. The French delegation did promise that the Jewish settlements would have free use of the waters of the Upper Jordan and the Yarmouk, although they would remain in French hands (Ra'anan 1955, p. 136). The Litani was excluded from this arrangement. Article 8 of the Franco-British Convention, therefore, included a call for a joint committee to examine the irrigation and hydroelectric potential of the Upper Jordan and Yarmouk, 'after the needs of the territories under French Mandate', and added that,

In connection with this examination the French government will give its representatives the most liberal instructions for the employment of the surplus of these waters for the benefit of Palestine (cited in Hof 1985, p. 14).

Although the location of water resources had been an important, sometimes overriding, issue with some of the actors involved in determining the boundaries of these territories, it is clear in the outcome that other issues took precedence over the need for unified water basin development. These other factors ranged from the geo-strategic—the location of roads and oil pipelines—to political alliances and relation-ships between British, French, Jews and Arabs, to how well-versed one or another negotiator was in biblical geography. The final boundaries were the result of competing needs and abilities of each of the people and entities involved in the negotiations. Because of limited land and resources, no two political entities could achieve all of their economic, historic and strategic requirements.

Between the wars, water became the focus of the greater political argument over how to develop the budding states in the region, and what the 'economic absorptive capacity' would be for immigration. Plans included the well-known Ionides (1939) and Lowdermilk (1944) plans for the Jordan River, and developments included the first gauging

---

[1] 'You will flounder if you like, but you will not flounder at our expense'.

station on the Tigris and an irrigation project for 65,000 hectares north of Baghdad.

Along the Nile, meanwhile, developments were moving in the direction of cooperation, at least between Egypt and the Sudan. In 1925, a new commission, made up of a representative each of the Egyptian and British governments, and a Dutch engineer, made recommendations based on the 1920 estimates which would lead finally to the Nile Water Agreement of 1929 (Krishna in Starr and Stoll 1988, p. 25). An amount of 4000 MCM/year was allocated to the Sudan but the entire timely flow (from 20 January to 15 July) and a total annual amount of 48,000 MCM/year was reserved for Egypt. This agreement allowed for two dams, the Sennar Dam and the Jebel Auliya Dam, to be constructed on the White Nile as envisaged in the Century Storage Scheme, although Egypt reserved the right for on-site inspection and a veto over any additional construction which might affect its own water resources (Shahin 1986).

B. 1945–1964: Unilateral Development and Sporadic Negotiations

As the borders of the new states of the Middle East were defined in the 1940s and 1950s, each country began to develop its own water resources unilaterally. On the Jordan, the legacy of the Mandates and the 1948 war was a river divided in a manner in which conflict over water resource development was inevitable. However, the shooting that did break out in the 1950s led to two years of some of the most intense negotiations ever between Arabs and Israelis—the Johnston negotiations. Unilateral development likewise led to negotiated agreements on the Nile.

Even as the dust was settling from the 1948 war, Syria approached Israel with a secret offer which, for the first time, linked three topics which would define the negotiating issues for the coming decades— peace, refugee resettlement, and water. Colonel Hosni Zaim took control of Syria in a U.S.-sponsored military coup in April 1949, with a promise that he would do 'something constructive' about the Arab-Israeli problem. That month, he sent a secret message to Israeli Prime Minister David Ben-Gurion offering to sign a separate peace agreement, establish a joint militia, and settle 300,000 Palestinian refugees in Syrian territory, in exchange for some 'minor border changes' along the cease-fire line and half of the Sea of Galilee (Shalev 1989). Ben-Gurion was reluctant to make such an agreement and signed a limited armistice instead. Less than a year later, Zaim was overthrown.

In 1951, several states announced unilateral plans for the Jordan watershed. Arab states began to discuss organized exploitation of two northern sources of the Jordan—the Hasbani and the Banias (Stevens 1965, p. 38). The Israelis made public their All-Israel Plan, based on James Hays' idea of a 'TVA on the Jordan', which in turn was based on the Lowdermilk proposals. The All-Israel Plan included the draining of Huleh Lake and swamps, diversion of the northern Jordan River and construction of a carrier to the coastal plain and Negev Desert—the first out-of-basin transfer for the watershed (Naff and Matson 1984, p. 35).

Jordan announced a plan to irrigate the East Ghor of the Jordan Valley by tapping the Yarmouk (Stevens 1965, p. 39). At Jordan's announcement, Israel closed the gates of an existing dam south of the Sea of Galilee and began draining the Huleh swamps, which lay within the demilitarized zone with Syria. These actions led to a series of border skirmishes between Israel and Syria which escalated over the summer of 1951 and prompted Israeli Foreign Minister Moshe Sharrett to declare clearly that, 'Our soldiers in the north are defending the Jordan water sources so that water may be brought to the farmers of the Negev' (Stevens 1965, p. 39).

In March 1953, Jordan and the U.N. Relief and Works Agency for Palestine Refugees (UNRWA) signed an agreement to begin implementing the Bunger Plan which called for a dam at Maqarin on the Yarmouk River with a storage capacity of 480 MCM, and a diversion dam at Addasiyah which would direct gravity flow along the East Ghor of the Jordan Valley. The water would open land for irrigation, provide power for Syria and Jordan, and offer resettlement for 100,000 Palestinian refugees. In June 1953, Jordan and Syria agreed to share the Yarmouk but Israel protested that its riparian rights were not being recognized (Naff and Matson 1984, p. 38).

In July 1953, Israel began construction on the intake of its National Water Carrier at Gesher B'not Ya'akov, north of the Sea of Galilee and in the demilitarized zone. Syria deployed its armed forces along the border (Davis et al. 1980, pp. 3, 8) and artillery units opened fire on the construction and engineering sites (Cooley 1984, pp. 3, 10). Syria also protested to the U.N. and, though a 1954 resolution for the resumption of work by Israel carried a majority, the USSR vetoed the resolution. The Israelis then moved the intake to its current site at Eshed Kinrot on the north-western shore of the Sea of Galilee (Garbell 1965, p. 30).

This was a doubly costly move for Israel. First, water salinity is much higher in the lake than in the upper Jordan. The initial water pumped in

1964 was actually unsuitable for some types of agriculture. Since that time, Israel has diverted saline springs away from the lake and filtered carrier water through artificial recharge to ease this problem (Stevens 1965, p. 9). Second, the water from B'not Ya'akov would have flowed to the Negev by gravity alone. Instead, 450 MCM/year is currently pumped a height of 250 m before it starts its 240-km journey southward. Israel currently uses 20 per cent of its overall energy requirements to move water from one place to another (State of Israel 1988, p. 136).

Against this tense background, President Dwight Eisenhower sent his special envoy Eric Johnston to the Mid-east in October 1953 to try to mediate a comprehensive settlement of the Jordan River system allocations (Main 1953). Johnston's initial proposals were based on a study carried out by Charles Main and the TVA at the request of UNRWA to develop the area's water resources and to provide for refugee resettlement. The TVA addressed the problem with the regional approach Lowdermilk had advocated a decade earlier. As Gordon Clapp, chairman of the TVA, wrote in his letter of presentation, 'the report describes the elements of an efficient arrangement of water supply within the watershed of the Jordan River System. It does not consider political factors or attempt to set this system into the national boundaries now prevailing' (Main 1953). This apolitical, basin-wide approach produced not only the thorough technical report which was to be the basis of two years of negotiations, but also stunning oversize maps which delineate only one border—that of the Jordan River watershed.

The Main Plan had, of course, other motives on the part of the United States, and advantages other than the technical details:

The plan, designed to tempt the Arabs into at least limited cooperation with the Israelis, was a third-rate idea with at least a second-rate chance of success because it had a first-rate negotiator, Eric Johnston, to advocate it. Its only advantage was that it made sense (Copeland 1969, p. 109).

The major features of the Main Plan included small dams on the Hasbani, Dan and Banias, a medium size (175 MCM storage) dam at Maqarin, additional storage in the Sea of Galilee, and gravity-flow canals down both sides of the Jordan Valley. The Main Plan excluded the Litani and described only in-basin use of the Jordan River water, although it conceded that, 'it is recognized that each of these countries may have different ideas about the specific areas within their boundaries to which these waters might be directed' (Main 1953). Preliminary allocations gave Israel 394 MCM/year, Jordan 774 MCM/year, and Syria 45 MCM/year.

Israel responded to the Main proposal with the Cotton Plan which incorporated many of Lowdermilk's ideas. This plan called for inclusion of the Litani, out-of-basin transfers to the coastal plain and the Negev, and the use of the Sea of Galilee as the main storage facility, thereby diluting its salinity. It allocated Israel 1290 MCM/year, including 400 MCM/year from the Litani, Jordan 575 MCM/year, Syria 30 MCM/year and Lebanon 450 MCM/year.

In 1954, representatives from Lebanon, Syria, Jordan and Egypt established the Arab League Technical Committee under Egyptian leadership and formulated the 'Arab Plan'. It reaffirmed in-basin use, rejected storage in the Sea of Galilee, which lies wholly in Israel, and excluded the Litani. The Arab representatives also objected to the refugee resettlement as a goal. The Arab Plan's principal difference from the Main Plan was in the water allocated to each state. Israel was to receive 182 MCM/year, Jordan 698 MCM/year, Syria 132 MCM/year, and Lebanon 35 MCM/year in addition to keeping all of the Litani.

Johnston worked until the end of 1955 to reconcile these proposals in a Unified Plan amenable to all of the states involved. His dealings were bolstered by a U.S. offer to fund two-thirds of the development costs, and given a boost when a land survey of Jordan suggested that that country needed less water for its future needs than was previously thought.

Johnston addressed the objections of both sides, and accomplished no small degree of compromise, although his neglect of groundwater issues would later prove an important oversight. Though they had not met face to face for these negotiations, all states agreed on the need for a regional approach. Israel gave up on integration of the Litani, and the Arabs agreed to allow out-of-basin transfer. The Arabs at first objected, but finally agreed, to storage at both the Maqarin Dam and the Sea of Galilee so long as neither side would have physical control over the share available to the other. Israel objected, but finally agreed, to international supervision of withdrawals and construction. Allocations under the Unified Plan, later known as the Johnston Plan, included 400 MCM/year to Israel, 720 MCM/year to Jordan, 132 MCM/year to Syria and 35 MCM/year to Lebanon (unpublished summaries, U.S. Department of State 1955, 1956).

The technical committees from both sides accepted the Unified Plan, and the Israeli Cabinet approved it without vote in July 1955. President Nasser of Egypt became an active advocate because Johnston's proposals seemed to deal with the Arab-Israeli conflict and the Palestinian problem simultaneously. Among other proposals, Johnston envisioned the diversion of Nile water to the western Sinai Desert to resettle two million Palestinian refugees.

Despite the forward momentum, the Arab League Council decided not to accept the plan in October 1955, and the momentum died out. In a 1955 letter lobbying against acceptance of the plan, the Arab Higher Committee for Palestine explained part of the underlying reluctance to enter into agreement:

The scheme is another step made by imperialists and Zionists to attain their ends, territorial expansion in the heart of the Arab homeland, under the attractive guise of 'economic interests' (cited in Medzini ed. 1976, p. 487).

Although the agreement was never ratified, both sides have generally adhered to the technical details and allocations even while proceeding with unilateral development. Agreement was encouraged by the United States, which promised funding for future water development projects only as long as the Johnston allocations were adhered to (Wishart 1990). From that time to the present, Israeli and Jordanian water officials have met two or three times a year at so-called 'Picnic Table Talks' at the confluence of the Jordan and Yarmouk Rivers to discuss flow rates and allocations.

To the west at the same time, hydrologic developments were caught up in the nationalist movements of both Egypt and the Sudan. The Aswan High dam, with a projected storage capacity of 156,000 MCM/year, was proposed in 1952 by the new Egyptian government, but debate over whether it was to be built as a unilateral Egyptian project or as a cooperative project with the Sudan was kept out of negotiations until 1954. The negotiations which ensued, and which were carried out with the Sudan's struggle for independence as a back-drop, focused not only on what each country's legitimate allocation would be, but even whether the dam was the most efficient method of harnessing the waters of the Nile.

The Sudan, hoping to guarantee its own future water needs, advocated the planning envisioned in the Century Storage Scheme, with its emphasis on upstream controls, and balked at Egyptian desires for water additional to its 1929 allocations. Egypt justified these new allocations of 62,000 MCM/year of the river's projected net discharge of 70,000

MCM/year, on the basis of the 'primary needs' of its larger population and on the lack of any other water supplies. Negotiations were broken off and relations threatened to degenerate into military confrontation in 1958 when Egypt sent an unsuccessful expedition into the territory in dispute between the two countries (Naff and Matson 1984, pp. 145–7; Lowi 1990, p. 128).

The Sudan attained independence in 1956, but it was with the military regime which gained power in 1958 that Egypt adopted a more conciliatory tone in the negotiations which resumed, and with which the Nile Water Treaty was signed on 8 November 1959. The treaty estimated Egyptian water use at 48,000 MCM/year and that of the Sudan at 6000 MCM/year. The estimated increase in yield to result from the High Dam, 22,000 MCM/year, was divided between Egypt (7500 MCM/year) and the Sudan (14,500 MCM/year). The total recognized water rights for each state of 55,000 MCM/year for Egypt and 18,500 MCM/year for the Sudan are still upheld today. The treaty also provided for a Sudanese water 'loan' to Egypt of up to 1500 MCM/year through 1977; for a Permanent Joint Technical Committee to resolve disputes and jointly review claims by any other riparian; and for equal sharing of future increases in the yield of the Nile (Whitington and Haynes 1985; Krishna in Starr and Stoll 1988, pp. 28–30). Egypt and the Sudan agreed that the combined needs of other riparians, at the time still under British rule, would not exceed 1000–2000 MCM/year. Ethiopia served notice in 1957 that it would pursue unilateral development of the Nile water resources within its territory, estimated at 75–85 per cent of the annual flow, and suggestions were made recently that Ethiopia may eventually claim up to 40,000 MCM/year for its irrigation needs both within and outside the Nile watershed (Jovanovic 1985, p. 85). No other state riparian to the Nile has ever exercised a legal claim to the waters allocated in the 1959 treaty (Whitington and McClelland 1992, p. 145).

C. 1964–1980s: 'Water Wars' and Territorial Adjustments

As each state developed its water resources unilaterally, their plans began to overlap. On the Jordan, the resulting tensions helped lead to a cycle of conflict which, exacerbated by other disputes, ended in war in 1967. Likewise on the Euphrates, water-related tensions led to the Syrian-Iraq 'water crisis' in 1974–75.

A 1963 agreement between Jordanian King Hussein and Ya'akov Herzog, envoy of Israeli Prime Minister Levi Eshkol, had spelled out an agreement on the allocation of the Jordan River water in return for Israeli acquiescence to U.S. tank sales to Jordan (Kershner 1990, p. 11). By 1964, Israel had completed enough of its National Water Carrier, so that actual diversions from the Jordan River basin to the coastal plain and the Negev were imminent. Although Jordan was also about to begin extracting Yarmuk water for its East Ghor Canal, it was the Israeli diversion which promoted President Nasser to call for the First Arab Summit in January 1964, including heads of state from the region and North Africa, specifically to discuss a joint strategy on water.

The options presented to the Summit were to complain to the U.N., divert the upper Jordan tributaries into Arab states, as had been discussed by Syria and Jordan since 1953, or to go to war (Schmida 1983, p. 19). The decision to divert the rivers prevailed at a Second Summit in September 1964, and the states agreed to finance a Headwater Diversion project in Lebanon and Syria and to help Jordan build a dam on the Yarmouk. They also made tentative military plans to defend the diversion project (Shemesh 1988, p. 38).

In 1964, Israel began withdrawing 320 MCM/year of Jordan river water for its National Water Carrier, and Jordan completed a major phase of its East Ghor Canal (Inbar and Maos 1984, p. 21). In 1965, the Arab states began construction of their Headwater Diversion Plan to prevent the Jordan headwaters from reaching Israel. The plan was to divert the Hasbani into the Litani in Lebanon and the Banias into the Yarmouk where it would be impounded for Jordan and Syria by a dam at Mukheiba. The diversion would take away up to 125 MCM/year, cut by 35 per cent the installed capacity of the Israeli Carrier, and increase the salinity in the Sea of Galilee by 60 ppm (United States Central Intelligence Agency 1962; Inbar and Maos 1984, p. 22; Naff and Matson 1984, p. 43).

Although a 1964 U.S. State Department memorandum concluded that the Arab Diversion seemed, 'unlikely to cause large-scale hostilities' (U.S. Department of State memorandum 1964), Israel declared the impending diversion as an 'infringement of its sovereign rights' (Naff and Matson 1984, p. 44). To a visiting U.S. delegation, Israeli Prime Minister Levi Eshkol declared that, 'Israel was not trigger-happy, but if it came to it, we would have to fight for our waters' (U.S. Department of State memorandum 1965).

The U.S. had supported the Israeli Water Carrier within the Johnston allocations and had both opposed the All-Arab Diversion and expressed doubt that it would be completed—Lebanon had stopped work on the diversion project in July 1965 (Hof 1985, p. 36). It was made clear to

Israel, though, that the U.S., 'would oppose you if you take preemptive action' (U.S. Department of State memorandum 1965). Nevertheless, in March, May and August 1965, the Israeli army attacked the diversion works in Syria.

These events set off what has been called 'a prolonged chain reaction of border violence that linked directly to the events that led to the (June 1967) war' (Professor Nadav Safran cited in Cooley 1984, p. 16). Border incidents continued between Israel and Syria, finally triggering air battles in July 1966 and April 1967.

Even as tensions were leading to the following week's outbreak of the 1967 War, the U.S. Departments of Interior and State convened an 'International Conference on Water for Peace' in Washington D.C. during 23–31 May 1967. Building on advances in nuclear energy and the possibility of inexpensive nuclear desalination, President Johnson had, in 1965, announced a 'massive, cooperative, international effort to find solutions for Man's water problems', which he dubbed 'the Water-for-Peace Program' (cited in Skolnikoff 1967, p. 157). The 1967 Conference had 6400 participants from 94 countries, including Israel, Egypt, Jordan, Yemen and Saudi Arabia (United States Departments of Interior and State 1967).

In the territorial gains and improvements in geo-strategic positioning which Israel achieved in the war which broke out the following week, Israel also improved its 'hydrostrategic' position. (See Figure 2—International Borders, 1967–Present.) With the Golan Heights, it now held all of the headwaters of the Jordan, with the exception of a section of the Hasbani, which, together with a view over much of the Yarmouk, made the Headwater Diversion impossible. The Mukheiba Dam was destroyed and the Maqarin Dam abandoned. And the West Bank not only provided riparian access to the entire length of the Jordan River, but it overlay three major aquifers, two of which Israel had been tapping into from its side of the Green Line since 1955 (Garbell 1965, p. 30). Jordan had planned to transport 70–150 MCM/year from the Yarmouk River to the West Bank. These plans, too, were abandoned.

In the wake of the 1967 war, former President Eisenhower who, ten years earlier had sent Eric Johnston to the Mid-east to negotiate a regional water plan, made public a new cooperation scheme which he, former Atomic Energy Commissioner Lewis Strauss, and Alvin Weinberg, director of the Oak Ridge National Laboratories, had formulated, which they called simply, 'A Proposal for Our Time'. Their plan called for three nuclear desalination plants—one each on the

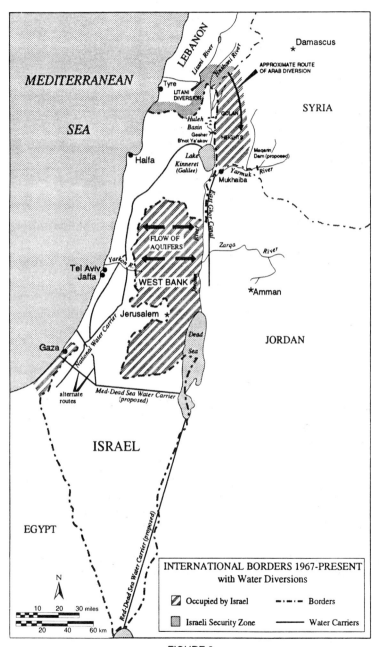

FIGURE 2

Mediterranean coast in Egypt and Israel, and one on the Gulf of Aqaba in Jordan—producing a combined output of about 1400 MCM of fresh water a year—roughly the usable flow of the entire Jordan River—as well as 'an enormous amount' of electric power. (Oak Ridge National Laboratories, Summary Report 1971; Strauss 1967).

Recently declassified documents show that an additional site was considered, at Gaza (Oak Ridge National Laboratories, Gaza Area 1970). At this site, a major consideration was the possibility of refugee resettlement, although sections of the report dealing with that aspect were excised from declassification.

As Eisenhower saw it, the availability of these new sources of energy and water would make possible entire 'agro-industrial complexes', making an additional 4500 km^2 of barren land arable, and providing work and agriculture to help settle more than a million Arab refugees (Eisenhower 1968). The project, which would cost about a billion (1967) U.S. dollars, would be funded by an international corporation set up for the purpose, and be supervised by the International Atomic Energy Agency. Moreover, Eisenhower predicted that,

. . . the collaboration of Arab and Jew in a practical and profitable enterprise of this magnitude might well be the first, long step toward a permanent peace (Eisenhower 1968, p. 77).

The project was studied in detail over the course of the next five years by a technical group made up of Arabs, Israelis, and Americans centred at the Oak Ridge National Laboratories. Although joint U.S.-Israeli studies on nuclear desalination dating back to 1964 had looked promising (U.S. Department of State memorandum, 14 December 1977), the 'Proposal for Our Time' eventually faltered on economic grounds, along with the dangers of introducing nuclear technology to the region, but the effort was finally called off because of political resistance. Nevertheless, two years of cooperative research in Oak Ridge, Tennessee, along with lessons learned during the Johnston negotiations twelve years earlier, showed that, on the technical level at least, cooperation over regional water resources and planning was possible. The Agro-Industrial Complex, which was to be the last attempt at regionwide water cooperation, was finally shelved in the early 1970s.

During the war between Israel and the combined forces of Egypt and Syria in 1973, water played only an incidental strategic role. Touring the Golan Heights with then Water Commissioner Menahem Cantor in the autumn of 1973, Defence Minister Moshe Dayan expressed concern that Israel's development of small-scale dams on the Golan Heights was proceeding so slowly. Dayan saw the strategic potential of these dams as tank barricades against Syrian forces. Citing budget limitations, Cantor was given encouragement and budget to proceed more quickly. Dayan was scheduled to tour the sites again on Sunday, 7 October, but the war broke out on the previous day. It is unclear how the dams finally performed their strategic function (interview, Menahem Cantor, November 1991).

Along the Euphrates, however, unilateral developments came very close to ending in warfare the following year. The three riparians to the river—Turkey, Syria and Iraq—had coexisted with varying degrees of hydropolitical tension throughout the 1960s. At that time, population pressures drove unilateral developments, particularly in southern Anatolia with the Keban Dam (1965–73), and in Syria with the Tabqa Dam (1968–73) (Lowi 1990, p. 108).

Bilateral and tripartite meetings, occasionally with Soviet involvement, had been carried out between the three riparians since the mid-1960s, although no formal agreements had been reached by the time the Keban and Tabqa Dams began to fill late in 1973, resulting in decreased flow downstream. In mid-1974, Syria agreed to an Iraqi request that Syria allow an additional flow of 200 MCM/year from Tabqa. The following year, however, the Iraqis claimed that the flow had been dropped from the normal 920 m³/s to an 'intolerable' 197 m³/s, and asked that the Arab League intervene. The Syrians claimed that less than half the river's normal flow had reached *its* borders that year and, after a barrage of mutually hostile statements, pulled out of an Arab League technical committee formed to mediate in the conflict. By the end of May 1975, hydropolitical relations between Iraq and Syria, exacerbated by other political differences, threatened to turn violent. Syria closed its airspace to Iraqi flights and and both Syria and Iraq reportedly transferred troops to their mutual border. Only mediation on the part of Saudi Arabia was able to break the increasing tension, and on 3 June the parties arrived at an agreement which averted the impending violence. Although the terms of the agreement were not made public, Naff and Matson (1984, p. 94) cite Iraqi sources as privately stating that the agreement called for Syria to keep 40 per cent of the flow of the Euphrates within its borders, and to allow the remaining 60 per cent through to Iraq.

Other negotiations proved less fruitful along the Jordan. In 1977,

Jordanian water officials approached their Israeli counterparts through U.S. intermediaries and requested a high-level meeting to discuss rebuilding the low dam at Mukheiba on the Yarmouk, the northern side of which would have abutted Israeli territory. One meeting was held that year in a Zurich hotel with three ministerial-level representatives from each side present. Israeli representatives expressed approval of the dam, one side of which would abut on Israeli territory—a more even year-round flow would benefit both sides—and agreed to further discussion on this and other regional water planning issues (unpublished minutes, 6 May 1977). In elections that year, however, the Israeli government shifted from Labour- to Likud-led for the first time, and the new ministers did not pursue the dialogue with the Jordanians. Direct ministerial negotiations were not held again on water issues except for a brief meeting in Jericho in 1985, although the 'Picnic Table Talks', on allocations of the Yarmouk River, continued at the technical level.

Meanwhile, tensions were being somewhat reduced along other borders. In 1978, Egypt and Israel signed the Camp David peace accords—the first between Israel and an Arab country. At a September 1979 meeting with the Israeli Press, President Anwar Sadat discussed plans for a pipeline to bring Nile water to the recently returned Sinai Peninsula. 'Once we bring [it] to Sinai,' he asked, 'why should we not bring some of this water to the Negev?' (Spector and Gruen 1980, p. 10.) The offer was reiterated and elaborated upon in discussions with Prime Minister Menachem Begin in 1981. Israel would be provided with 365 MCM/year in exchange for 'solution of the Palestinian problem and the liberation of Jerusalem' (Krishna in Starr and Stoll eds. 1988, p. 32).

The offer was immediately rejected by almost all parties concerned. Prime Minister Begin objected to the quid pro quo, stressing that Israel would not trade its sovereignty over a unified Jerusalem for economic gain. Nationalists on both sides were also opposed to the idea—Egyptians did not want to share this vital resource with Israel, and Israelis did not like the idea of being vulnerable to upstream control. Israeli Agriculture Minister Ariel Sharon was quoted as saying, 'I would hate to be in a situation in which the Egyptians could close our taps whenever they wished' (Spector and Gruen 1980, p. 10).

The strongest opposition to the offer came from another region entirely. Ethiopia, 2500 kilometres upriver, charged Egypt with misusing its share of Nile water. In a sharp retort, President Sadat warned against Ethiopian action:

We do not need permission from Ethiopia or the Soviet Union to divert our Nile water . . . If Ethiopia takes any action to block the Nile waters, there will be no alternative for us but to use force. Tampering with the rights of a nation to water is tampering with its life and a decision to go to war on this score is indisputable in the international community (Krishna in Starr and Stoll eds. 1988, pp. 33–34).

President Sadat was assassinated in 1981. Although technical and economic details of a Nile River diversion have since been developed (see, for example, Kally in Fishelson ed. 1989; Dinar and Wolf 1993), the plan was never implemented except for a small irrigation diversion into the western Sinai.

D. Israel, the West Bank, and Gaza

Ever since the 1973 war, the regional conflict focus has shifted from being Israeli-Arab to Israeli-Palestinian. This is true regarding water conflicts, as well. In fact, while earlier periods were marked by major water projects and regionwide water conflicts, this most recent period has mostly been one of internal adjustments within each state to optimize existing water resources. Israeli water policy, however, also includes territory and populations under military occupation, whose final status has yet to be determined. Because of the hydrography of these areas, the focus has also shifted from a surface water to a groundwater conflict.

As mentioned earlier, Israel took control of the West Bank and Gaza in 1967, including the recharge areas for aquifers which flow west and north-west from the West Bank into Israel, and east to the Jordan Valley (Kahan 1987, p. 21). The entire renewable recharge of these first two aquifers is already being exploited and the recharge of the third is close to being depleted as well. The annual 'safe yield' and current use of these aquifers is given in the table.

	Yield (MCM/year)	Consumption (MCM/year)	
		Israel	Palestinians
Western aquifer	320	300	20
Eastern aquifer	125	25	50
North-east aquifer	140	120	20

The total consumption within the West Bank is 35 MCM/year, mostly from wells, for Israeli settlements, and 115 MCM/year, from wells and cisterns, for Palestinians. In Gaza, the natural annual recharge of 60

MCM/year is routinely augmented with an additional 35 MCM/year in groundwater overdraft, resulting in increasing salination from saltwater intrusion.

In twenty-four years of occupation, a growing West Bank and Gaza population, along with burgeoning Jewish settlements, has increased the burden on the limited groundwater supply, resulting in an exacerbation of already tense political relations. Palestinians have objected strenuously to Israeli control of local water resources and to settlement development, which they see as being at their territorial and hydrologic expense (see, for example, Davis 1980; Dillman 1989; Zarour and Isaac 1992).

In 1967, Israel nationalized all West Bank and Gaza water and limits were placed on the amount withdrawn from each existing well. Since that time, the only permits for new Palestinian wells which have been granted are for domestic needs. Agricultural usage was capped at 1968 levels and all subsequent extension of land under irrigation has been through increased efficiency (Richardson 1984). At the same time, seventeen wells were drilled in the West Bank to provide water to the new Israeli settlements. Some Palestinian wells were undercut and desiccated, notably at al-Auja and Bardala, because of the deeper, more powerful Israeli wells (Dillman 1989, pp. 56–7). Of the 47 MCM/year pumped in the mountain area, 14 MCM/year, or 30 per cent, goes to the Jewish settlements. The eastern aquifer, which flows into the Jordan Valley, is the only one not being overexploited, but Palestinians have not been allowed to expand their water resources in this region either (Dillman 1989, p. 57).

Israelis argue that Palestinian agriculture can expand using water saved through more efficient agricultural practices. For example, modern methods of irrigation have helped Palestinian farmers in the Jitflik valley increase vegetable production tenfold without significantly increasing water needs (Rymon and Or 1989). They argue further that any limits imposed on pumping have depended on the situation of each aquifer at the time the permit was requested—not on whether the applicants were Arabs or Jews—and that, with only one exception, desiccated Palestinian wells have been supplied with alternative sources (Info. Briefing 1986; interviews, Golani, October 1991; Shmuel Cantor, December 1991).

Israeli authorities viewed these actions as defensive, of a sort. Hydrogeologically, Israel is down-gradient of the West Bank aquifers. In essence, groundwater flows, albeit extremely slowly, from the recharge areas and upland aquifers of the West Bank down to those on the Israeli side of the Green Line on its way to the sea. Israel had been tapping up to

270 MCM/year of this groundwater from its side of the Green Line since 1955 (Garbell 1965, p. 30). Any uncontrolled, extensive groundwater development in the newly occupied territories would threaten these coastal wells with salt-water intrusion from the sea, causing serious damage (Jaffee Center 1989, p. 200).

With about 30 per cent of Israeli water originating on the West Bank, the Israelis are aware of the need to limit groundwater exploitation in these territories in order to protect both the resources themselves, and the wells from salt-water intrusion. To this end, they have even imported surface water from the National Water Carrier to the Ramallah and Hebron hill region for Arab domestic use rather than allowing additional drilling (Spector and Gruen 1980, p. 10). Further, four or five Israeli settlements built in the late 1970s around Elkanna, near the Green Line, may have been sited to guarantee continued Israeli control of some of the contested water (State of Israel memoranda June–July 1977; Pedhatzor 1989).

Palestinians have objected to this increasing control and integration into the Israeli grid. Legal arguments often refer, at least in part, to the Fourth Geneva Convention's discussion of territories under military occupation (see, for example, Dillman 1989; El-Hindi 1990). In principle, it is argued, the resources of occupied territory cannot be exported for the benefit of the occupying power. Israeli authorities reject these arguments, usually claiming that the Convention is not applicable to the West Bank or Gaza because the powers these territories were wrested from were not, themselves, legitimate rulers (Blum 1968; El-Hindi 1990). Egypt was itself a military occupier of Gaza and only Britain and Pakistan recognized Jordan's 1950 annexation of the West Bank. Also, it is pointed out that the water Israel uses is not being exported but rather flows naturally seaward, and, because Israel has been pumping that water since 1955, it has 'prior appropriation' ('first in time, first in right') rights to the water.

Eventually, the final political and hydrographic status of this region will have to be determined. Aside from politics or nationalisms, hydrologic reasoning would seem to dictate that this determination be done sooner rather than later. As one U.N. report notes,

The present integration of the basic water services in the occupied territories with those of Israel is about to lead to the complete dependence of the former services on those of Israel and will eventually make the separation of the two very costly and difficult (cited in Dillman 1989, p. 63).

E. 1980s to the Present: Hydrologic Limits and Peace-Making

By the mid-1980s, each of the countries riparian to the rivers of the Middle East began to approach its hydrologic limits, and the potential for either conflict or cooperation took on new urgency, both in the region and abroad.

The fundamental tenet of ecological systems is, 'Everything is connected to everything else' (Holling 1978, p. 26). For those dependent on a watershed approaching the limits of available water one might add: 'Everything you do will affect someone else'. As the riparians to the Nile, Jordan and Euphrates watersheds began to run out of hydrologic room to manoeuvre, this tenet became increasingly apparent.

In 1983, construction on the Sudanese Jonglei Canal was halted because of that country's civil war. This project, which had been discussed since 1904, was projected to add 18,000 MCM/year to the yield of the Nile. This increase would have been shared equally between the Sudan and Egypt, according to the terms of the 1959 Nile Water Agreement (Whitington and McClellan 1992).

In 1985, plans for a deep well near Herodian in the West Bank were made public. Funded by an American fundamentalist Christian group, this project would have brought 18 MCM/year to both Arabs and Jews on the West Bank. Wary that the size and depth of the project might undercut their wells, some Palestinians had international pressure brought to bear on the Israelis and Americans involved, and the project was halted (Caponera 1991).

Meanwhile, the Syrians, who had lost access to the Banias springs in 1967, began a series of small impoundment dams on the headwaters of the Yarmouk in their territory in the late 1970s. By August 1988, twenty dams were in place with combined capacity of 156 MCM/year (Sofer and Kliot 1988, p. 19). That capacity has since grown to 27 dams with a combined storage of about 250 MCM/year (Gruen 1991, p. 24; interview, Shmuel Cantor, December 1991). According to Gruen (1991, p. 24), the Syrians have plans to expand this storage to 366 MCM/year by 2010. These Syrian impoundments are in contradiction with their 1953 agreement with Jordan, which allocates seven-eighths of the water of the Yarmuk to Jordan in exchange for two-thirds of the hydro-power from the planned Maqarin dam (Caponera 1991, p. 10).

Because the Maqarin, or Unity, Dam was never built, winter run-off, most of which Jordan cannot now capture for use in its East Ghor Canal, flows almost unimpeded downstream to Israel. This situation has allowed Israel to use more than the 25 MCM/year allocated to it from the Yarmuk by the Johnston accords.

In secret negotiations during 1989–90, mediated by the U.S. State Department's Richard Armitage, Israel has argued that it has prior appropriation rights to a greater share of Yarmouk water—to between 40 and 100 MCM/year—because of its greater use over the years, and because of its new responsibilities to the West Bank (Gruen 1991; Kolars 1992). Agreement on this issue is a prerequisite to building the Unity Dam. The World Bank has agreed to help finance the project only if all of the riparians agree to the technical details.

Along the Euphrates, development of the Turkish GAP project in southern Anatolia urged the riparians to address their differences. A 1987 visit to Damascus by Turkish Prime Minister Turgut Ozal reportedly resulted in a signed agreement for the Turks to guarantee a minimum flow of 500 m³/s across the border with Syria. According to Kolars and Mitchell (1991, p. 286), this total of 16,000 MCM/year is in accordance with prior Syrian requests. However, according to Naff and Matson (1984), this is also the amount that Iraq insisted on in 1967, leaving a potential shortfall. A tripartite meeting between Turkish, Syrian and Iraqi ministers was held in November 1986, but yielded few results (Kolars and Mitchell 1991).

Talks between the three countries were held again in January 1990, when Turkey closed the gates to the reservoir on the Ataturk Dam, essentially shutting off the flow of the Euphrates for 30 days. At this meeting, Iraq again insisted that a flow of 500 m³/s cross the Syrian-Iraqi border. The Turkish representatives responded that this was a technical issue rather than one of politics and the meetings stalled. The Gulf War which broke out later that month precluded additional negotiations (Kolars and Mitchell 1991, pp. 288–9).

With tensions developing during the 1980s, the United States, which had initiated both the Johnston negotiations in the 1950s and the water-for-peace process during the 1960s, became convinced anew of water's potential for conflict. By the end of the 1980s, comprehensive studies on the strategic aspects of water in the Mid-east and the potential for conflict had been conducted by the U.S. Defense Intelligence Agency— Naff and Matson (1984); the Center for Strategic and International Studies—Starr and Stoll (1988); and the Israeli Foreign Ministry—Sofer and Kliot (1988); also, the House of Representatives subcommittee on Europe and the Middle East had held a hearing of Middle East water issues (June 1990). Each concluded not only that the water resources of the region held great potential for conflict, but that, of the Middle East water basins, the Jordan presented the most likely flashpoint.

In the thinking of the Defense Intelligence Agency:

Water ignores artificial political boundaries; in an undeveloped environment it flows according to the terrain. When man—in order to make better use of water for himself—changes the natural distribution system, he also changes traditional use patterns. This can be extremely disruptive and upsetting to other riparian users. The result is often political conflict if not outright military action. Military factors are often the *de-facto* determinants in resolving riparian relationships in the Middle East (personal communication, 3 July 1991).

By 1991, several events combined to shift the emphasis on the potential for 'hydro-conflict' in the Middle East watershed to the potential for 'hydro-cooperation'.

The first event was natural, but limited to the Jordan basin. Three years of below-average rainfall caused a dramatic tightening in the water management practices of each of the riparians, including rationing, cut-backs to agriculture by as much 30 per cent, and restructuring of water pricing and allocations. Although these steps placed short-term hardships on those affected, they also showed that, for years of normal rainfall, there was still some flexibility in the system. Most water decision-makers agree that these steps, particularly regarding pricing practices and allocations to agriculture, were long overdue.

The next series of events were geo-political and regionwide in nature. The Gulf War in 1990 and the collapse of the Soviet Union caused a re-alignment of political alliances in the Middle East which finally made possible the first public face-to-face peace talks between Arabs and Israelis, in Madrid on 30 October 1991.

With countries still in the throes of drought, water was mentioned as a motivating factor for the talks. Jordan, as has been mentioned, is squeezed hydrologically between two neighbours attempting to re-interpret prior agreements, but otherwise has no major territorial disputes with Israel. A researcher at the Middle East Studies Center in Amman therefore suggested that, 'Jordan is being pushed to the peace talks because of water' (interview, Mohammed Ma'ali, November 1991). Mohammed Beni Hani, the head of Jordan's water authority, is one of Jordan's twelve delegates to the peace talks.

At the opening ceremonies in Madrid, Dr Haider Abdel-Shafi, the head of the Palestinian delegation, included in his opening remarks a call for 'the return of Palestinian land and its life-giving waters'.

During the bilateral negotiations between Israel and each of its neighbours, it was agreed that a second track be established for multilateral negotiations on five subjects deemed 'regional'. These subjects included ecology, energy, economic cooperation, arms reduction and *water resources*.

With the opening of the peace talks, the emphasis in international arenas quickly went from the potential for conflict over water to its potential as a vehicle for cooperation. Seminars and conferences were held throughout 1990 and 1991 in the U.S., Canada, Europe and the Middle East on the possibilities for cooperation over water resources. The World Bank held a seminar on the topic, as did the U.S. Department of State, and the Center for Foreign Affairs. Increasingly, both Arab and Israeli academics and policymakers have taken part together in these conferences.

In Jerusalem, the Israel/Palestine Center for Research and Information (IPCRI) began holding round-table discussions and simulated negotiations on water in December 1990. In December 1992, IPCRI cosponsored, with the Hebrew University of Jerusalem and the Applied Research Institute in Bethlehem, the 'First Israeli/Palestinian International Conference on Water' in Zurich.

On a larger scale, the first round of multilateral negotiations on water was held in Vienna in May 1992, with representatives from more than twenty countries participating. At the meeting, each part agreed to compile a programme for regional development, which will then be examined in the United States for any commonalities which could be exploited to induce cooperation. The same approach is being taken by the World Bank, which commissioned similar studies from the states in the region. In conjunction with the peace talks, less-public and less-official dialogue, called the 'Track 2 talks', have been held between Israelis and Arabs in the U.S.A.

These breakthroughs in water talks may have repercussions on negotiations on other topics as well. In the words of Munther Haddadin, a Jordanian delegate, 'Water seems to be leading the Peace Talks.'

CONCLUSION

The waters of the Middle East have been the focus both of bitter conflict over a scarce and vital resource, and of cooperation even between otherwise hostile neighbours. From the Nile to the Jordan to the Euphrates, armies have been mobilized and treaties signed over this precious commodity. In recent years, the needs of ever-increasing populations and burgeoning national development have begun to approach and sometimes exceed local hydrologic limits. As shortages become more acute, unilateral plans increasingly impose on co-riparians, physically driving home the potential hazards of resource confict—or the benefits of regional cooperation.

As we determine whether the future will take the shape of increasing riparian disputes and perhaps armed hostilities, or, alternatively, of greater cooperation and regionwide planning, some observations from the turbulent hydropolitical history of the region may be in order, along with their implications for the future of the region:

1. Observation: The single overriding impediment to regional water resources planning is the lack of a basin-wide water authority on any of the river systems under discussion.

Implication: Political objections to cooperation must be overcome to achieve efficient management of these river systems. Any negotiations should emphasize regional planning as a crucial goal.

2. Observation: The link between water resources and political alternatives is inextricable, with water scarcity leading directly to both heightened political tensions and opportunities for cooperation.

Implication: For negotiations for a political settlement to be successful, they will have to also address solutions to the water conflict. Likewise, workable solutions to the problems of regional water shortage should also address to constraints posed by regional politics.

3. Observation: Water has historically been a factor in Middle East population distribution, including some border considerations.

Implication: Successful negotiations over, for example, Jewish immigration or Palestinian 'right of return' will have to incorporate the hydrologic limitations of the region.

4. Observation: No dispute between Arabs and Israelis, on water or on any other issue, has ever been resolved without third-party (usually U.S.) sponsorship and active participation.

and

5. Observation: The better a state's 'hydro-strategic' position, the less interest it has in reaching a water-sharing agreement.

Implication: Strong third-party involvement will be necessary for successful negotiations. The U.S., or other sponsors of negotiations, should be prepared with a comprehensive strategy to induce cooperation, with particular emphasis on the upstream riparians.

6. Observation: Projects of limited and implicit cooperation have been successful even in advance of political solutions between the parties involved (e.g. Picnic Table talks, water-for-peace process). Nevertheless, explicit cooperation (e.g. Maqarin Dam), has not preceded political relations.

and

7. *Observation:* The more complex a proposal is technically, the more complex it is politically.

Implication: In the context of regional talks, progress in negotiations over water resources may encourage dialogue on other, more contentious, issues. While water continues to 'lead' the peace talks, projects to induce cooperation can be designed in a step-wise fashion beginning with 'small and doable', and leading to ever-increasing integration, always remaining on the cutting edge of political relations.

8. *Observation:* The two conditions at the core of political viability of watersharing are *equity* of the agreement or project (that is, how much does each participant get), and *control* by each party of its own primary water sources (or, from where does it come, and whose hand is on the tap).

Implication: These two contentious issues will have to be addressed fairly early in negotiations. Unless a water-sharing agreement is worked out, with each party having its historic as well as future needs addressed, any negotiations over intricate cooperative projects will be building on accumulated ill-will.

If emphasis is placed on easing regional water tensions, some breathing space might be gained, allowing for more complex political and historical difficulties to be negotiated. In fact, because the water problems to be solved involve all of the parties at conflict, and because these issues are so fundamental, the search for regional solutions may actually be used as a tool to facilitate cooperation.

The peace talks of the 1990s have included the mutual impact of water on political decision-making. Seventy years of regional water development, however, have both heightened the political stakes of water issues, and left less hydrologic room for manoeuvrability. Given, though, that an important political precedent has been set in Madrid—public face-to-face negotiations, the lack of which has precluded explicit cooperation in the past—and given the lessons learned through 100 years of 'hydro-diplomacy', a new potential for regional planning and cooperation may have been reached. One can hope that, after 100 years, the lessons have been learned.

As one American involved in the water-for-peace process of the 1960s is quoted as having said, 'Water is an eloquent advocate for reason' (Strauss 1967).

ACKNOWLEDGEMENTS

The author wishes to thank the Center for Environment Policy Studies, University of Wisconsin-Madison; the U.S. Institute of Peace; and the United Nations University for their support during completion of this work.

BIBLIOGRAPHY

Beaumont, Peter. 'Transboundary Water Disputes in the Middle East'. Submitted at a conference on Transboundary Waters in the Middle East, Ankara, September 1991.

Blum, Y. 'The Missing Reversionary: Reflections on the Status of Judea and Samaria'. *Israeli Law Review*, Vol. 3, 1968.

Caponera, Dante A. 'Legal and Institutional Concepts of Cooperation'. Submitted at a conference on Transboundary Waters in the Middle East, Ankara, September 1991.

Cooley, John. 'The War Over Water'. *Foreign Policy*, Spring 1984, No. 54, pp. 3–26.

Copeland, Miles. *The Game of Nations*. New York: College Notes and Texts, 1969.

Davis, Uri, Antonia Maks, John Richardson. 'Israel's Water Policies'. *Journal of Palestine Studies*, Winter 1980, 9:2:34, pp. 3–32.

Dillman, Jeffrey. 'Water Rights in the Occupied Territories'. *Journal of Palestine Studies*, Autumn 1989, 19:1:73, pp. 46–71.

Dinar, Ariel and Aaron Wolf. 'International Markets for Water and the Potential for Regional Cooperation: The Case of the Western Middle East' in M. Shechter ed. *Sharing Scarce Fresh Water Resources in the Mediterranean Basin: An Economic Perspective*. Boston: Kluwer Academic, 1993 (forthcoming).

Eisenhower, Dwight. 'A Proposal for Our Time'. *Reader's Digest*, June 1968, pp. 75–80.

El-Hindi, Jamal L. 'Note, The West Bank Aquifer and Conventions Regarding Laws of Belligerent Occupations'. *Michigan Journal of International Law*, Summer 1990, 11:4, pp. 1400–23.

El-Yussif, Faruk. 'Condensed History of Water Resources Developments in Mesopotamia'. *Water International*, Vol. 8, pp. 19–22, 1983.

Esco Foundation. *Palestine: A Study of Jewish, Arab, and British Policies*. New Haven: Yale University Press, 1947.

Fishelson, Gideon, ed. *Economic Cooperation in the Middle East*. Boulder: Westview Press, 1989.

Frey, Frederick. 'The Political Context of Conflict and Cooperation Over International River Basins'. Prepared for a conference on Middle East Water Crisis, Waterloo, 1992.

Friedman, Isaiah ed. *The Rise of Israel* (A Facsimile Series Reproducing 1900 Documents in 39 Volumes). New York: Garland, 1987.

Fromkin, David. *A Peace to End All Peace: The Fall of the Ottoman Empire and the Creation of the Modern Middle East*. New York: Avon, 1989.

Garbell, Maurice. 'The Jordan Valley Plan'. *Scientific American*, March 1965, 212:3, pp. 23–31.

Gruen, George. *The Water Crisis: The Next Middle East Crisis?* Los Angeles: Wiesenthal Center, 1991.

Hof, Frederic. *Galilee Divided: The Israel-Lebanon Frontier, 1916–1984*. Boulder: Westview, 1985.

Holling, C. 'Model Invalidation and Belief'. Chapter 7 in *Adapting Environmental Assessment and Management*. New York: John Wiley and Sons, 1978.

Inbar, Moshe and Jacob Maos. 'Water Resource Planning and Development in the Northern Jordan Valley'. *Water International*, 1984, Vol. 9, pp. 18–25.

Ingrams, Doreen. *Palestine Papers, 1917–1922*. London: Murray, 1972.

Issar, Arie. *Water Shall Flow from the Rock: Hydrogeology and Climate in the Lands of the Bible*. New York: Springer-Verlag, 1990.

Jaffee Center for Strategic Studies. *The West Bank and Gaza: Israel's Options for Peace*. Tel Aviv: Tel Aviv University, 1989.

Jovanovic, D. 'Ethiopian Interests in the Division of the Nile River Waters'. *Water International*, Vol. 10, pp. 82–5, 1985.

Kahan, David. *Agriculture and Water Resources in the West Bank and Gaza (1967–1987)*. Jerusalem: The Jerusalem Post, 1987.

Kally, Elisha. *The Struggle for Water*. Tel Aviv: Hakibbutz Hameuhad Publishing, 1965 (Hebrew).

Kershner, Isabel. 'Talking Water'. *Jerusalem Report*, 25 October 1990.

Kolars, John. 'Trickle of Hope: Negotiating Water Rights is Critical to Peace in the Middle East'. *The Sciences*, November/December 1992.

Kolars, John and William Mitchell. *The Euphrates River and the Southeast Anatolia Development Project*. Carbondale and Edwardsville: Southern Illinois University Press, 1991.

Lowdermilk, Walter. *Palestine: Land of Promise*. New York: Harper and Bros., 1944.

Lowi, Miriam. *The Politics of Water Under Conditions of Scarcity and Conflict: The Jordan River and Riparian States*. Ph.D. Dissertation, Princeton University, 1990.

McCarthy, Justin. *The Population of Palestine: Population Statistics of the Late Ottoman Period and the Mandate*. New York: Columbia University Press, 1990.

Main, Chas. T., Inc. *The Unified Development of the Water Resources of the Jordan Valley Region*. Knoxville: Tennessee Valley Authority, 1953.

Medzini, Meron, ed. *Israel's Foreign Relations*. Jerusalem: Ministry for Foreign Affairs, 1976.

Naff, Thomas and Ruth Matson, eds. *Water in the Middle East: Conflict or Cooperation?* Boulder: Westview Press, 1984.

Oak Ridge National Laboratories. 'Middle East Study Application of Large Water-Producing Energy Centers: Gaza Area Development and Refugee Resettlement'. Draft, 10 November 1970.

———, 'Middle East Study Application of Large Water-Producing Energy Centers: Summary'. Draft, 15 November 1971.

Pedhatzor, Reuven. 'A Shared Tap'. *Ha'aretz*. 3 May 1989 (Hebrew).

Ra'anan, Uri. *The Frontiers of a Nation: A Re-examination of the Forces which Created the Palestine Mandate and Determined its Territorial Shape*. Westport (CT): Hyperion Press, 1955.

Reisner, Marc. *Cadillac Desert: The American West and Its Disappearing Water*. New York: Viking, 1986.

Richardson, John. *The West Bank: A Portrait*. Washington: The Middle East Institute, 1984.

Rymon, Dan and Uri Or. 'Advanced Technologies in Traditional Agriculture, A Case Study: Drip Fertigation in the Jiftlik Valley'. Unpublished, 1989.

Sachar, Howard. *The Emergence of the Middle East: 1914–1924*. New York: Knopf, 1969.

Sachar, Howard. *A History of Israel*. New York: Knopf, 1979 (Vol. I), 1987 (Vol. II).

Schmida, Leslie. *Keys to Control: Israel's Pursuit of Arab Water Resources*. American Educational Trust, 1983.

Shahin, Mamdouh. *Hydrology of the Nile Basin*. New York: Elsevier, 1986.

Shalev, Ayre. *Cooperation in the Shadow of Conflict*. (Hebrew) Tel Aviv: Ma'arachot, 1989.

Shemesh, Moshe. *The Palestinian Entity 1959–1974: Arab Politics and the PLO*. London: Frank Cass, 1988.

Skolnikoff, Eugene. *Science, Technology and American Foreign Policy*. Boston: MIT Press, 1967.

Sofer, Arnon and Nurit Kliot. 'Regional Water Resources'. Israel Ministry of Defense, 1988 (Hebrew).

Spector, Lea and George Gruen. 'Waters of Controversy: Implications for the Arab-Israel Peace Process'. *American Jewish Committee*, December 1980, pp. 1–11.

Starr, Joyce and Daniel Stoll, eds. *The Politics of Scarcity: Water in the Middle East*. Boulder: Westview Press, 1988.

State of Israel. 'Minutes of Meeting, 6 May 1977'. Unpublished.

———, Environmental Protection Service. *The Environment in Israel*. 4th edition, Jerusalem, 1988.

———, Foreign Ministry. 'Memorandum on Water in the Golan Heights, Judea and Samaria, and the Jordan Valley'. Unpublished, 7 March 1977 (Hebrew).

———, Israel State Archives. *Documents on the Foreign Policy of Israel*. Jerusalem, 1981.

———, Ministry of Agriculture. 'Water Security for the State of Israel Today and in the Future'. Unpublished memorandum, Minister of Agriculture Katz-Oz to Prime Minister Yitzhak Shamir, 14 May 1989 (Hebrew).

———, Ministry of Agriculture. *Hydrological Year-Book of Israel: 1986–1988*. Jerusalem, 1990.

———, State Comptroller. *Report on Water Management in Israel*. Jerusalem, 1990 (Hebrew).

Stevens, Georgiana. *Jordan River Partition*. Stanford: The Hoover Institution, 1965.

Strauss, Lewis. 'Dwight Eisenhower's "Proposal for Our Time" '. *National Review*, September 1967, 19:37, pp. 1008–11.

Tuchman, Barbara. *Bible and Sword*. New York: Ballantine, 1956.

United States, Army Corps of Engineers. *Water in the Sand: A Survey of Middle East Water Issues*. Draft, 1991.

———, Central Intelligence Agency. 'Struggle for Jordan Waters'. Unpublished memorandum, May 1962.

———, Departments of Interior and State. *Water for Peace*. Proceedings from International Conference on Water for Peace, Washington, D.C., 23–31 May 1967. Eight Volumes.

———, Department of State. 'The Jordan Valley Plan (Summary of Johnston Negotiations)'. Unpublished, 30 September 1955; revised 31 January 1956.

———, Department of State. 'Memorandum for the President of Visit of Israeli Prime Minister Eshkol'. Unpublished, 22 May 1964.

———, Department of State. 'Harriman Mission to Israel'. Unpublished Memorandum of Conversation, 25–27 February 1965.

Waterbury, John. *Hydropolitics of the Nile Valley*. New York: Syracuse University Press, 1979.

Weizmann, Chaim. *The Letters and Papers of Chaim Weizmann*. Leonard Stein ed. London: Oxford University Press, 1968.

Whitington, Dale and Kingsley Haynes. 'Nile Water for Whom? Emerging Conflicts in Water Allocation for Agricultural Expansion in Egypt and Sudan', in Beaumont, P. and K. McLachlan eds. *Agricultural Development in the Middle East*. New York: John Wiley and Sons, 1985.

Whitington, Dale and Elizabeth McClelland. 'Opportunities for Regional and International Cooperation in the Nile Basin'. *Water International*, 1992, Vol.17, pp. 144–54.

Wishart, David. 'The Breakdown of the Johnston Negotiations over the Jordan Waters'. *Middle Eastern Studies*. 26:4, October 1990.

Wolf, Aaron. 'The Impact of Scarce Water Resources on the Arab-Israeli Conflict: An Interdisciplinary Study of Water Conflict Analysis and Proposals for Conflict Resolution'. Ph.D. Dissertation, University of Wisconsin—Madison, 1992.

Zarour, Hisham and Jad Isaac. 'Nature's Apportionment and the Open Market: A Promising Solution Convergence to the Arab-Israeli Water Conflict'. Submitted to a conference on The Middle East Water Crisis, Waterloo, 7–9 May 1992.

3 / Problems of International River Management: The Case of the Euphrates

JOHN KOLARS

Sheer size does not determine the challenge that a river presents to those who would utilize its water wisely. The Euphrates River has only one-third the volume of the Nile, and even when combined with its unruly sister, the Tigris, in the Shatt al-Arab (Map 1), the flow of the two streams is less than that of Egypt's source of life. Nevertheless, the use of the Euphrates' waters is as ancient, as complex, and still as critical today as that of the waters of the Nile, the Indus, or the Huang He.

This analysis examines the development, management, and the present and predicted use of the Euphrates River. A similar treatment of the Tigris River must remain for the future, although the role which the latter stream may play vis-à-vis the development of the Euphrates is considered herein. The discussion begins with a model of Middle-Eastern river use which offers insight into the development of the Euphrates-Tigris River system. A physical description of the streams follows. The Turkish South-east Anatolia Development Project (Turkish acronym: GAP) and its impact on the ecology and economies of the entire basin is the focus of attention. The riparian activities, needs and expectations of Syria and Iraq are presented in less detail, largely because Syria must react to Turkish moves rather than act independently, and because in the case of Iraq, not much can be determined at present about the state of river management. The impact of the twin rivers' future upon the Arabian/ Persian Gulf (hereafter referred to as the Gulf) is briefly mentioned. A consideration of the problems and potential of the Euphrates' current and future management concludes the discussion.

A Model of Middle-Eastern River Basin Development

Concern over equitable sharing of international river resources in the Middle East has resulted in theoretical models of riparian partners' behaviour at the supra-national level (Naff and Matson 1984, pp. 181–97) although no attempt has been made to translate such ideas into managerial terms. By the same token, little work has been done regarding the intramural relationship between political, economic, and technological

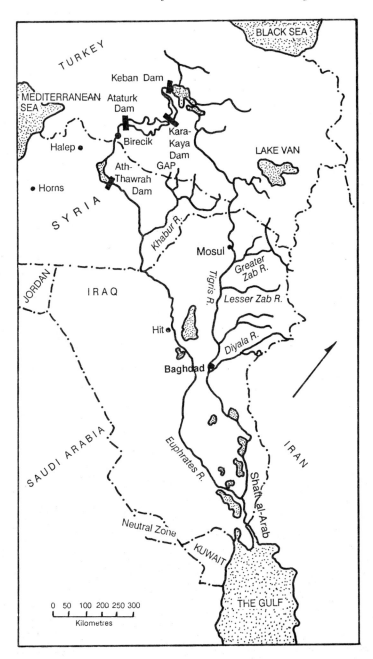

Map 1. The Tigris and Euphrates River System (sketch map).

processes accompanying river basin development *within* Middle Eastern Countries.[1] The discussion which follows attempts to describe such activities and to assess them as they relate to the Euphrates-Tigris system. While emphasis is placed upon the Euphrates portion of the GAP in Turkey, Syria's development efforts vis-à-vis its General Administration for the Development of the Euphrates Basin (GADEB) are also considered. Less can be said of Iraq's current use and management of the two rivers, although an attempt is made to assess the impact of the upper two riparians' activities on the latter country.

The first observation that can be made concerning river development and management in the Middle East is that the process begins with holistic visions of what might be accomplished by 'taming' a river, but with little thought given to consequences and problems. Such grand ideas are followed eventually by linear activities whereby structures (i.e. dams, reservoirs, irrigation systems) are put in place, only to have holistic commentators, in turn, question the impact of such projects upon society and the environment. Dreamers and social critics, engineers and politicians, engage in monologues which seldom are heard or modified by the others' points of view. The result is the prioritizing of internal needs and actions which beget intra-mural competition leading to inefficient or failed projects. And, as the results of upstream projects resonate downstream, international misunderstandings follow.

A second commonality shared by Middle-Eastern rivers is that each rises in a mountainous catchment area. In the case of the Euphrates, 98 per cent of the flow of that stream originates in the highlands of Eastern Turkey.[2] The Tigris River receives 38 per cent of its water directly from Turkey and approximately another 11 per cent from tributaries which also rise there (see Kolars 1992a for a review of water availability in the Middle East). (The Nile's sources are more complex with the floods of the Blue Nile, the Sobat, and the Atbara from the Ethiopian highlands providing, according to Waterbury [1979, p. 23], 95 per cent of the waters reaching Egypt, while a scant 5 per cent comes from the equatorial lakes of Africa. Evans states that 84 per cent of the flow at Aswan originates from Ethiopia [1990, pp. 20–21].) In every case, the source

---

[1] Other rivers have received such treatment. Weatherford and Brown (1983), explore such an approach in depth.

[2] Syria's share of the Euphrates is usually given as 12 per cent of the total. Analysis by this author shows that as much as 10 per cent out of that 12 per cent originates from the Syrian springs of the Khabur and Balikh streams which have their catchments north of the border in Turkey where pumping of groundwater could diminish or staunch their flow.

areas fall outside the borders of downstream riparian states and have been the last to be developed.

A third feature of these rivers is that they are *exotic* in character. That is, they receive all their waters near their sources and grow smaller as they flow to the sea. The Euphrates receives no additional water south of Deir ez-Zor and its confluence with the Khabur.[3] (The Nile receives none north of the mouth of the Atbara.)

Fourth, these streams and all other streams in the Middle East and similar arid to semi-arid regions of the world have extremely high seasonal and multi-annual variance in their flow. The Euphrates' annual flow in Turkey at Birecik near the Syrian border ranged from 42.7 billion m^3 in 1963 to 15.3 billion m^3 in 1961 (Kolars and Mitchell 1991, table 5.10). Ephemeral peak flows recorded at Hit, Iraq, during the period 1924–73 were as high as 7390 m^3/sec (1969) and as low as 850 m^3/s (1930) (Kolars and Mitchell 1991, pp. 90–97). (According to Waterbury [1979, pp. 22–23], while the average annual discharge of the Nile at Aswan from 1900 to 1959 was 84 billion m^3, 'The standard deviation from that mean was about 20 billion m^3 annually'. Moreover, 'More than 80 per cent of the river's total discharge occurs from August to October while nearly 20 per cent is spread over the remaining nine months'.)

The net result of the conditions described above is that the management of such rivers is extremely difficult even if they are confined within the borders of a single nation. When, as in the case of the Middle East, several riparian states are involved (three for the Euphrates, nine for the Nile) the problems are multiplied.

This is particularly true when the history of land use along the rivers is considered. In almost every case, exotic rivers and their flood plains have been settled and utilized first in their lower reaches. Upstream areas (Ethiopia, the interior of Africa, the highlands of eastern Turkey) have been the last to be developed, and the rivers in the latter areas and at later times are often used for hydroelectric generation as well as agriculture.

Another way of considering these uses is that in early historic times irrigated agriculture served local needs, but with the growth of nationalism and with crop production geared to the developing world market in the 20th century new national and international pressures have been placed on limited supplies of water. This, in turn, has been exacerbated by a rapid increase in population, particularly in downstream nations.

---

[3] Some exchange of water occurs in Iraq between the Tigris and the Euphrates Rivers and might be considered a downstream addition to the latter, but this is an irregular, unpredictable occurrence.

Gamal Hamdan (Waterbury 1979, pp. 25–42) describes the development of the Nile valley in terms of four periods of technology: the *Geotechnic* wherein natural basins on the flood plain were used for agriculture, the *Paleotechnic* in which basin irrigation became fully developed, the *Neotechnic* during which valley agriculture was placed on a perennial basis, and the *Biotechnic* which necessitates an annual supply of 'timely' water, that is, water reliably available during the long period between the annual floods and able to override any multi-annual variance due to changes in precipitation in source areas.

Both Egypt and Iraq (lower Mesopotamia) fully experienced the first two of these states as does southern Sudan today. South-eastern Turkey and Ethiopia, remaining as they have both isolated and underdeveloped until the second half of this century, have taken little advantage of the rivers until recently. The same may be said of Syria, where pumping from the Euphrates by private farmers in order to grow cotton began only in the 1950s (Hinnebusch 1989, Chapter 8, and Sanlaville and Metral 1979, pp. 229–40). However, all three riparians on the Euphrates, as well as the Sudan and Egypt on the Nile, are engaged in or rapidly becoming involved with neotechnic and biotechnic agricultural practices.

Among the largest of these is Turkey's South-east Anatolia Development Project (Turkish acronym: GAP). The GAP incorporates the construction of 21 dams and 19 hydropower plants on the Euphrates and Tigris Rivers. One million hectares of land are scheduled to be irrigated with water from the former stream and 625,000 ha from the latter. The GAP will have a total of 7500 MW installed capacity with an average annual production of 26 billion kWh. This in turn represents 19 per cent of the 8.5 million ha of the economically irrigable land in Turkey, and 20.5 per cent of the country's hydropower.[4]

---

[4] The analyst can become entangled in numbers when considering statements such as this. These figures are routinely given for descriptions of the GAP. In almost every case, the Keban Dam, upstream, the second largest on the river, is omitted because it technically falls outside of the GAP, although its management is an integral part of the management of the Euphrates River.

Furthermore, Table 5.2 of the *Final Master Plan* (Vol. 2, p. 5.26) lists nine hydropower plants scheduled or in action in the Euphrates system (excluding the Keban Dam but including one run-of-the-river plant), and 14 such installations (including four run-of-the-river plants) on the Tigris. It also lists 18 active reservoirs on the Euphrates system (including the Keban Dam) and 11 on the Tigris.

Figures vary from publication to publication where such broad estimates are concerned. For example, the *Final Master Plan Report* in its *Executive Summary* (p. 2) gives a value of 22 per cent and 118 billion kWh of 'economically viable hydropwer potential'.

The emphasis upon the production of cash crops has become intertwined with the need to produce hydroelectric energy using the same waters. This situation, as will be seen, has the potential for internal conflict leading to dissonance over the sharing of international streams. This situation occurred on the Euphrates in 1974 and again in 1981. In the first case, when the Turks' need for additional energy led to their building the Keban Dam (Kolars 1986), its reservoir was filled by chance at the same time that Lake Assad behind the Tabqa (ath-Thawrah) Dam was being filled for similar reasons, i.e. Syrian hopes of expanding irrigated agriculture and generating needed power. Unknown to either nation, this coincided with one of the driest years in decades. As a result the flow into Iraq was reduced to a trickle and only intervention by Saudi Arabia prevented open hostilities between Iraq and Syria. On the second occasion, it was thought by the interim Turkish government (after the 1980 army coup) that the upcoming elections and acceptance of a new Turkish constitution would be eased by an abundance of consumer electricity. To ensure this, the Keban Reservoir was run down to a low point, and during its refilling, downstream shortages produced international reverberations although the situation did not reach crisis proportions thanks to an improvement in weather conditions.

This situation was repeated with more intensity in 1990 when the flow of the river was interrupted for 27 days in order to partially fill the Ataturk Reservoir. As the Euphrates downstream flow diminished, Iraq and Syria formed an uneasy detente in order to attend tripartite ministerial meetings held in Ankara regarding the crisis. At those meetings, the Turks stated that the matter was a technical one; the Iraqis and the Syrians insisted that it was political. At that point, the ministers returned home. Shortly thereafter the invasion of Kuwait took place and the situation was put on hold. (See Chalabi and Majzoub 1993 for a thorough review of these events.)

From the Turkish point of view, the stakes have been raised, for the Ataturk Dam is not only intended to produce large amounts of electricity (see below), but also as many as a million hectares of land are to be irrigated in the south-east with water from its reservoir. This is expected to generate large amounts of foreign exchange from the sale of new crops. High priority is also given to the potential that GAP has for raising the standard of living in south-east Anatolia, which in turn is seen as a way to ameliorate the discontent of the local Kurds who form a majority in the region.

Thus, the Turkish dream to harness the waters of the Euphrates

River—one shared and contested by both Premier Suleyman Demirel and President Turgut Ozal—in order to allay Turkey's serious energy shortage has become translated into a matter of prestige for the two political parties involved. Also, it has become a matter of internal security vis-à-vis the Kurdish problem, as well as a source of international confrontations with downstream partners.

Much the same thing happened in Syria. Early schemes to develop as many as 650,000 hectares along the Euphrates by building the ath-Thawrah Dam were reduced by 1983 to developing 345,000 ha and subsequently 240,000 ha. Inaccurate soil surveys conducted by German firms failed to warn the Syrians about the effect of gypsiferous soils both on canals and on field applications of water. The Rasafah Project originally estimated by the Russians to encompass 150,000 ha was actually abandoned, and no more than 208,000 ha (12,000 ha government projects; 196,000 ha private lands) were under irrigation in the Euphrates valley in 1985–86 (Kolars and Mitchell 1991, p. 274–82). Moreover, large tracts of fertile valley land have been lost beneath the waters of Lake Assad and to poor drainage and salination. Revisions in Syrian agricultural plans now place greater emphasis on dry farming and ancillary projects on the Khabur. Nevertheless, in the words of Raymond Hinnebusch:

While reduced priority to the Euphrates might make economic sense, the regime prestige invested in it and the multitude of bureaucratic interests at stake makes diversion of resources elsewhere politically unpalatable to decision-makers (1989, p. 220).

The situation on the Euphrates River, as elsewhere in the Middle East, finds geotechnic developments replacing earlier types of river utilization. These have taken two forms: (1) demands for basin-wide cooperative efforts which until now have largely foundered on the rocks of conflicting interests among upstream and downstream nations; and (2) the construction of mammoth dams capable in combination of holding back two or more years' flow of their streams. Such dams in downstream locations give temporary relief to the societies which they serve, but without binding riparian agreements remain vulnerable to similar upstream projects. Given the latter case, complications can be so intimately tied to internal problems at the state level that international negotiations between riparians may stall or even fail. In other words, *riverine foreign policy has to date been driven by* **domestic** *state-society relations*. It is within this context that the future of the Euphrates river must be considered.

The Euphrates-Tigris River System

The Euphrates and Tigris Rivers share a twin basin through their confluence near Basra to form the Shatt al-Arab in lower Iraq (Maps 2 and 3). The Shatt continues for 140 km to the Gulf, on the way being joined by the Karun, a major tributary, 32 km below Basra.[5] The combined annual, natural flow of the Shatt averages some 81.9 billion m³/s. However, removals and diversions as well as seasonal and multi-annual variation deny the usefulness of this figure.

The Euphrates River is the longest river (2700 km) in south-west Asia west of the Indus. It is formed in eastern Turkey by the confluence of the Karasu and the Murat Rivers 45 km north-west of Elazig. From that point it descends through the Anti-Taurus Mountains to the Syrian border south of Birecik, dropping an average 2 m/km. The observed average annual flow across the Turkish/Syrian border is 29.8 billion m³. The natural flow, prior to river development in Turkey has been estimated at 31.5 billion m³ (Kolars 1992c, p. 107). The flow of the Euphrates varies seasonally from a recorded minimum at Hit, Iraq, of 181 m³/s, to a maximum of 5200 m³/s at the same station.

After entering Syria, the river occupies an entrenched valley, flowing first south and then south-east into Iraq. Two tributaries which join the main stream from the left bank, the Balikh and Khabur, account for Syria's contribution to the flow of the river. These tributaries, however, receive most of their volume from springs immediately south of the Turkish/Syrian border, and have their catchments almost entirely inside of Turkey. Thus, their flow can be affected by the tapping of aquifers on the Turkish side. This author estimates that as much as 98 per cent of the Euphrates' waters therefore originate in Turkey, rather then the 88 per cent usually assumed.

No further water is added to the Euphrates downstream from the entry of the Khabur at Deir es-Zor, with the exception of irregular and infrequent hydrologic events in Iraq which may add some Tigris water to its flow. At Hit, located 360 km downstream from the Syrian border, the Iraqi portion of the Euphrates enters its alluvial plain. In the 735-km trip from Hit to the Gulf, the river drops only 53 m, and loses much of its waters in a series of natural and manmade distributaries. Far downstream near Nasiriya, the river becomes in part a tangle of channels draining into Lake Hammar, while the remainder finds its way to the Shatt.

---

[5] If an older confluence near Qurna is considered, the Shatt al-Arab is approximately 200 km in length. It should be noted that the Turks view the twin system as a single basin, while the Arab riparians view them as two separate basins (Chalabi and Majzoub 1992).

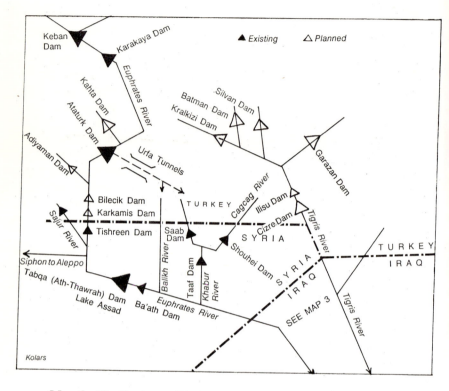

Map 2. The Euphrates-Tigris Basin: Existing and Planned
Developments (not to scale)

The Tigris River originates in south-eastern Turkey near Lake Hazar
and flows south-east to the Turkish city of Cizre whence it forms the
border between Syria and Turkey for 32 km before entering Iraq. The
Tigris reaches its alluvial plain midway between Tikrit and Samarra.
Unlike the Euphrates, this river receives water from numerous left-bank
tributaries which originate in the Zagros Mountains to the east. The
Greater Zab, the Lesser Zab, the Adhaim and the Diyala are the most
important of these streams and contribute approximately 28.7 billion m³
annually to the river, about 58 per cent of its natural flow at Qurna. The
main stream in Turkey and the Khabur River (not to be confused with the
Khabur shared by Turkey and Syria farther west) account for the
remaining annual flow of 20.5 billion m³.

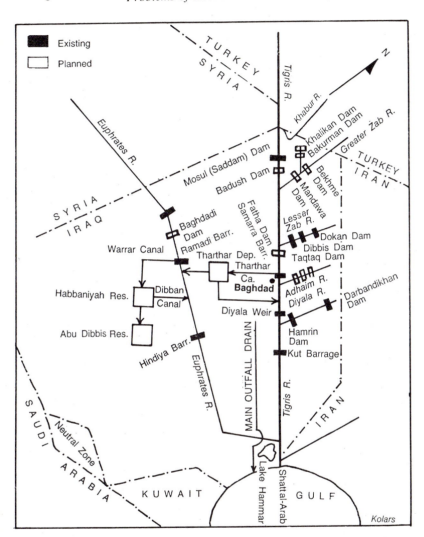

Map 3. River Development in Iraq: Schematic representation

The proximity of the Tigris' tributary sources in the Zagros Mountains accounts for wide variation in the volume of water carried by the river. When the spring snow-melt is accompanied by heavy rains, the Greater Zab may contribute 65 per cent of the river's volume in April and May. In addition to flood waters lost to distributaries farther downstream, high water is at times diverted from the Tigris into the Tharthar depression between it and the Euphrates to the west.[6] Thus, the flow of the main stream varies greatly along its length, as well as seasonally and from year to year. Near the confluence of the main stream with the Diyala the volume of flow may reach 14,000 m³/s, while downstream at Qurna the flow may be as low as 179 m³/s. The minimum flow recorded at Baghdad is 158 m³/s, the maximum 13,000 m³/s, while the mean is 1236 m³/s. Variation in the flow of both rivers ranges from conditions of severe drought to destructive flooding, and it is on this basis that the Turks make one of their strongest justifications for implementing the GAP with its giant dams and reservoirs capable of smoothing out such variance and providing a dependable year-round flow downstream. However, this argument has not been persuasive enough for the Syrians and Iraqis.

Turkey's South-west Anatolia Development Project is unquestionably the factor most responsible for creating change and producing contention regarding the use of the waters of the twin rivers. Iraq as the historical user might have tolerated Syria's newly developing demands, but the GAP through its magnitude has brought about a major confrontation between the three riparians who share the river. Though Turkey did not undertake the GAP in order to spite its neighbours or to intimidate them, its own energy, agricultural and foreign exchange needs required new vision and action which include the development of its major river resources. In any event, the GAP reverberates downstream in dramatic ways which make it necessary to understand Turkey's situation vis-à-vis such needs.

The Turkish Energy Base

To do this we must review Turkey's total energy picture, a salient feature of which is the rapidly escalating demand for electrical power. In the period between 1975 and 1982 Turkey's energy use increased by 30 per cent. At the same time, its total energy production increased only by 24 per cent. This situation, typical of a continuing trend, posed serious difficulties, for in those seven years petroleum consumption had increased by 18 per cent, and amounted to a total, annual import cost of nearly four billion dollars (Kolars 1986).

Even with subsequent price reductions, the petro-bill for Turkey amounted to 2.6 billion dollars in 1988 (Turkey 1990, table 293).

[6] The water can be released from there into the Euphrates, but extreme salinity in the Tharthar precludes such action except in extreme cases.

Persistent exploration for new oil deposits continues to prove unsuccessful, and Turkey's oil field at Batman, near Siirt, offers only limited supplies of poor quality. These fields, which are the northerly extension of the Arabian Gulf and Mosul deposits, are limited by extensive faulting and past volcanic activity, conditions common throughout Turkey. Thus, the petroleum situation described for 1982 has not changed for the better in the last ten years.

Turkish coal deposits at Zonguldak on the Black Sea are deep, contorted and dangerous, and expensive to mine. As of 1989, Turkey produced 1,973,000 tons of coal but consumed 4,687,000 tons, the balance of which was imported (Turkey 1991, table 114). While some natural gas is found in Turkish Thrace, imports from the Soviet Union in the same year cost 181 million dollars. In the area of lignite, asphaltite, geothermal energy, wood and animal fuel (*tezek*) Turkey consumes what it produces, but such sources, with the exception of lignite, are unimportant. Table 1 indicates the energy shares each source provided in 1989.

In the period 1982 through 1988, energy production from all sources increased by 33 per cent, but energy consumption as a percentage of production was 172 per cent in 1982 (Table 2). Coal production fell 18.8 per cent; lignite led all forms of domestic production, increasing by 85.7 per cent, but accounted for only 16 per cent of all Turkey's energy consumption in 1988 (Table 3). Mining lignite outstripped burning lignite by one per cent, perhaps indicating that the major thermal stations near Mugla and Kahraman-Maras were nearing capacity. Domestic natural gas remained insignificant (2.1 per cent of total energy produced) though its importation and use increased by over 1000 per cent (Table 3). Given the available energy sources, the picture thus presented is not encouraging save for the hydroelectric potential of the nation's rivers (Table 4).

The Hydro-power Potential of Turkey's Rivers

Despite the many claims made upon Turkey's rivers (see Table 5), there is an undeniable need for their being used to help offset Turkey's growing energy demands. In 1991, 61 hydro-projects installed in the country had a capacity of 7052 MW and produced 25,410 GWh (derived from Faralyali 1992, paras. 5 and 7, no page numbers). (See also Tables 6 and 7 of this paper).

The numerous rivers in Turkey offer a total economically viable hydro-power potential of 35,618 MW, with production, under average hydrological conditions, of 126,650 GWh. Not all of this will be realized immediately, although an additional 425 hydropower plants are under

Table 1. Production and Consumption of Turkish Energy by Source (1989)
Petroleum equivalents 1000s of tons

Coal	Lig.	Aspha	Nat. Gas	Petro.	Geoth.	Wood & Other	Animal & veg.	Hydro-elec.
PRODUCTION								
1973	10478	175	158	3020	35	5345	2580	4005
CONSUMPTION								
4687	10041	175	2878	22522	35	5345	2580	4005
PERCENTAGE OF TOTAL ENERGY CONSUMED								
9.0	19.1	0.3	5.5	43.2	+	10.2	4.9	7.0

(Imported electricity = 48 = 0.1%)

Source: Turkey 1991, table 114, p. 145.

Table 2. Changes in Primary Energy Production (1982 and 1988)
Petroleum equivalents 1000s of tons (10,000 kcal/kg)

Source	Production		% change	Consumption as a percentage of production in 1982
	1982	1988		
Total	21050	28089	33.4	172
Coal	2445	1986	−18.8	126
Lignite	4652	8638	85.7	99
Petroleum	2450	2692	9.9	691
Hydropower	3165	6467	204.0	100
Natural gas	41	90	220.0	100
Others (Asphalite, Wood, Waste, Geothermal, Elect. imports and 'Others')	8460	8210	−1.1	100

Source: Turkey, 1991 table 197, p. 225.

Table 3. Changes in Primary Energy Consumption (1982 and 1988)
Petroleum equivalents 1000s of tons (10,000 kcal/kg)

Source	Consumption		% change	Consumption as a percentage of production in 1988
	1982	1988		
Total*	36244	50758	40.0	181
Coal	3077	4606	49.7	232
Lignite	4616	8041	74.2	93
Petroleum	16924	22308	31.8	829
Hydropower	3165	6467	104.3	100
Natural gas	41	1062	2590.0	1180
Others (as in table 2)	8460	8248	−2.5	99.6

Source: Derived from: Turkey (1991, table 197, p. 225).
* Total energy consumption covers changes in stocks of secondary coal since 1982.

construction, programmed or planned. When completed, these will have an installed capacity of 22,358 MW and an average annual production of 80,460 GWh, 26 per cent of the estimated electricity that will be needed by Turkey in the year 2010. Key elements in this increase are dams and hydro-plants on the Euphrates and Tigris Rivers.

Although the amount of hydropower produced in 1989 amounted to only 7.7 per cent of the total energy used (including imported fuels), the potential energy offered by Turkey's rivers, in combination with new

Table 4. Installed Capacity by Establishment (1981 and 1988) 10^6 kWh (% change shown in parentheses)

Year	Total	TEK*	Chartered Companies	Auto Manufacturers	Municipal
1981					
Total	5537.6	4442.2	325.8	625.4	144.2
Hydro	2356.3	2097.5	219.8	12.2	26.8
Thermal	3181.3	2344.7	106.0	613.2	117.4
1988					
Total	14518.1 (262)	12981.5 (292)	378.4 (16.1)	1158.2 (185)	Phased out by 1983
Hydro	6218.3 (264)	5935.1 (283)	272.4 (23.9)	10.8 (−11.5)	—
Thermal	8299.8 (261)	7046.4 (301)	106.0 (0.0)	1147.4 (187)	—

Source: Derived from: Turkey (1990, table 198, p. 225).
* TEK — Turkish Electricity Authority.

Table 5. Turkish Stream and River Demarcated Boundaries

Neighbouring Country	Length of Shared Boundary	Wet Boundary	% Wet
Bulgaria	269	50	19
Greece	212	188	89
Iran	454	20	4
Iraq	331	38	11
Syria	877	76	9
Armenian Rep. (Former USSR)	610	243	40

Source: Bilen and Uskay (1991, table 3.1).

Table 6. Percentage Shares of Installed Generating Capacity by Establishment and Type (1982 and 1988)

Year	Total	TEK	Chartered Companies	Municipal
1982				
Share of Total	100.0	80.2	17.2	2.6
Hydro	42.6	47.2	67.5	18.6
Thermal	57.4	52.8	32.5	81.4
	100.0	100.0	100.0	
1988				
Share of total	100.0 ·	89.4	10.6	Phased out in 1983
Hydro	42.8	45.7	72.0	—
Thermal	57.2	54.3	28.0	—
	100.0	100.0	100.0	

Source: Derived from table 4.

opportunities for irrigated export crops and the foreign exchange they can earn, has been evident to Turkish planners for decades (Aydinelli 1940).

Turkish Electrical Production

In 1936 Turkey established an Electric Affairs Survey Administration which first considered the Keban Dam project on the Euphrates River. Further surveys in 1948 and the establishment of the State Irrigation

Table 7. Percentage Shares of Electricity Production by Establishment and Type (1982 and 1988)

Year	Total	TEK	Chartered Companies & Auto	Municipal	Imported
1982					
Share of total	100.0	78.4	15.3	0.2	6.1
Hydro	48.0	54.0	63.9	34.6	—
Thermal	45.9	46.0	36.1	65.4	—
Imports	6.1	—	—	—	100.0
	100.0				
1988					
Share of total	100.1	88.8	10.4	—	0.8
Hydro	59.8	63.8	70.2	—	—
Thermal	39.4	36.2	28.0	—	—
Imports	0.8	—	—	—	100.00
	100.0				

Source: Derived from: Turkey (1990, table 199, p. 226).

Works General Directorate in 1954 made clear the necessity of basin-wide planning. The State Hydraulic Works (DSI) and the State Planning Organization established a Euphrates Planning Office in Diyarbakir in 1961. This in turn led to the issuing of a viability report in 1963 and the awarding to an Italian-French firm the construction contract, in cooperation with the Turkish government, for the Keban Dam and Hydro-Electric Power Plant (HEPP) in 1968. The Keban HEPP became operational in 1975.

Meanwhile, parallel studies on the Tigris river were completed, and in 1977 development of the two rivers was subsumed under the title South-eastern Anatolia Project (GAP). Those early efforts were followed by more and more sophisticated projects, carried out entirely by Turkish engineers and Turkish construction companies. Coordination of the GAP was assigned to the State Planning Organization in 1986, and in November 1989 'the Development Administration for the South-eastern Anatolia Project was set up as a legal entity affiliated to the Prime Ministry' (EKA 1992).

In January 1992 Turkish installed electric generating capacity of all kinds reached 17,200 MW; at the same time gross consumption topped 60 billion kWh with a 9 per cent annual growth rate. Of the above capacity, 59 per cent was thermal and 41 per cent hydroelectric. Even so, average per capita consumption, according to Energy and Natural Resources Minister, Ersin Faralyali, was only 1051 kWh compared with the world per capita average of 2200 kWh (Faralyali 1992).

To rectify this situation, the gross per capita consumption level has been targeted to reach 2000 kWh by the end of the century. GAP is a key element in this effort. The centrepiece of GAP is the Ataturk Dam on the Euphrates River and its HEPP which will have an installed capacity of 2400 kW. Taken in conjunction with the Keban Dam and the Karakaya Dam and their associated HEPPs which are already on line, the three together will have a total installed capacity of 5530 MW producing 22.3 billion kWh annually.[7] In addition, by 1996 the Kralkizi, Dicle and Batman Dams and HEPPs will have been completed on the Tigris River, adding an additional 402 MW installed capacity with an estimated annual production of 1.6 GWh. Priority is aso being given to the construction of the necessary substations and EHV transmission lines to distribute this energy. Not only the GAP region will benefit

[7] The Keban Dam, while on the Euphrates River, is not considered to be part of the GAP. This discussion attempts wherever possible to include Keban data as part of the overall development of south-east Anatolia.

from such production, for 2177 km of national lines are being constructed or are under bid in order to more completely integrate the GAP with the national power network (Faralyali 1992).

Financing

While it is not the intention of this article to discuss the financing of Turkey's hydro-development programme in detail, the evolution of this issue over time is of note. The Keban Dam, farthest upstream, is a HEPP facility without agricultural water commitments.[8] The first to be built on the Euphrates, its financing met with little difficulty. Work, beginning in 1968, was financed by the European Investment Bank, USAID, and the French, German and Italian governments. The total cost in 1974 U.S. dollars, the year it was completed, was about $85 million.

In the years that followed, difficulty was experienced in finding international financing for the Karakaya Dam, next downstream and included in the GAP. Nevertheless a combination of Turkish and foreign funds was eventually found and the dam completed in 1988. By the time contracts were tendered to the Dogus Insaat ve Ticaret A.S. for the Ataturk Dam, questions over the downstream impact of the GAP precluded the loaning of funds by the World Bank for its construction. However, in March 1985 the Export-Import Bank of New York and the Manufacturers Hanover Trust loaned Turkey $111 million for the dam. European banks have also provided at least $460 million for equipment, but the bulk of the funding for the Ataturk Dam has been provided by the Turks themselves. As of mid-1992 the total cost of the Ataturk project had reached TL 11,241,000 million (1$ = 7600 TL on 15 October 1992) (EKA 1992).

The Birecik Dam, scheduled for construction downstream from the Ataturk, is being financed by a new, experimental scheme, Build, Operate and Transfer (BOT). The dam and power plant will be built by an international consortium (Birecik A.S., i.e. Birecik Incorporated Joint Stock Company) which will operate the facility until it has received a reasonable return on the stockholders' investments. Thereafter, the dam and HEPP will be transferred to the Turkish government.

All this activity has changed, enhanced and dramatized Turkey's role in the development of the Euphrates and Tigris Rivers. But this in turn has raised the question: what path will Turkish development of the twin

---

[8] Nevertheless, considerable irrigation is practised using streams *feeding* the reservoir. The amount of land involved is estimated to be between 30,000 and 50,000 ha.

rivers take? Will hydropower production complement or compete with the other goals that the GAP has established, and how will such decisions reflect downstream? An overview of Turkish water resources is necessary to answer this.

Turkey's Water Resources

Despite persistent water shortages in its major cities, and despite protests to the contrary by some of its officials, data presented by the Turkish State Hydraulic Works (Devlet Su Isleri 1984) indicate that with proper planning, adequate funding and reasonable reimbursement, the nation in the future should have water sufficient to farm all its lands suitable for irrigation, to slake the thirst of its cities, and have enough surplus to offer some help to its arid neighbours to the south.

Various challenges stand in the way of realizing such a glowing prediction. The geographical and temporal distribution of precipitation across Turkey is uneven. The western, central and south-eastern parts of the country tend to be drier then those in the north and north-east which receive copious amounts of rain. Turkey's major urban population concentrations and much of its fertile lands are found in water-deficit regions, a fact which complicates distribution for consumption. Moreover, precipitation varies dramatically from winter to summer, and has high seasonal and multi-annual variance.

Nevertheless, the twenty-six river basins within the country have a total average, annual runoff of 185 billion cubic metres, of which, the State Hydraulic Works (DSI) estimates, 62 billion m^3 will be consumed each year sometime after the year 2000 (Turkey 1984; Bilen and Uskay 1991, pp. 1, 2). Not all of the remaining 123 billion m^3 can be consumed, however, for rivers must provide hydro-electric power, and the tradeoff between energy and agriculture can at times be tricky. Much of the remaining water will be inaccessible or in quantities too small for economical recovery. Moreover, rivers must be kept flowing in order to sustain the environment and to meet downstream international obligations.

Turkey's Shared Water Resources

The nature of the above water resources is further complicated by the fact that of the 2763 km of border which Turkey shares with neighbouring countries, 615 are what may be termed 'wet' boundaries, that is, demarcated by a stream or river (Table 5). Turkey's success in negotiating agreements regarding the use of these boundary streams is noteworthy.

In 1927 Turkey and the Soviet Union signed a 'Treaty on the Beneficial Uses of Boundary Waters'. This treaty addressed the use of the Coruh, Kura, Arpa and Aras Rivers, the waters of which they agreed to share on a fifty-fifty basis. A Joint Boundary Water Commission was established (although without legal identity) to control the use of the frontier waters. In 1973 the two governments signed an additional Treaty on the Joint Construction of the Arpacay (Ahuryan) Storage Dam. After extensive feasibility studies, the dam was built and since 1986 has been operated by a joint technical commission. In 1973 the meanders of the Aras River were brought under control as far as possible through engineering measures, and the wet border it constitutes between Turkey and the former USSR was stabilized. (One might ponder, with the break-up of the Soviet Union, what future negotiations will take place between Turkey and the new Armenian Republic.)

In a similar vein, Turkey and Greece after the Treaty of Lausanne signed several protocols regarding the control and management of the Meric (Meritza) River which forms the boundary between Greek and Turkish Thrace. After 1951 the two governments with the assistance of the Harza Engineering firm of Chicago prepared a master plan including levees, flood control and drainage works in order to stabilize the border as well as to allow irrigation of 16,900 ha in Turkey and 11,600 ha in Greece. Flood control involving this scheme has already commenced. This plan, however, did not take the upstream sources of the Meric into account, and activities of the Bulgarian reach of the river have affected downstream activities. Nevertheless, the cooperation between Turkey and Greece on this issue is noteworthy.

In addition to the above shared border streams which the Turks choose to call *international rivers*, there are a number of other rivers which cross the borders of Turkey at an angle rather then forming mutual boundaries. Such streams are designated by the Turks as *transboundary rivers*.

The future of Turkey's *transboundary rivers* remains in large part to be decided. Here again, Turkey's role in Middle-Eastern affairs and development has changed sharply in the last two decades, and the importance of such rivers has been amplified through Turkey's undertaking the South-east Anatolia Development Project.

Protocol concerning the Euphrates and Tigris Rivers dates back to 1946 when Iraq and Turkey agreed that the rivers' control and management depended in great part upon the regulations of flow in Turkish source areas. Turkey, at that time, agreed to begin monitoring the two streams and to share related data with Iraq.

In 1980 Turkey and Iraq further specified the nature of the earlier protocol by establishing a Joint Technical Committee on Regional Waters. After a bilateral agreement in 1982, Syria joined the committee which subsequently held meetings in Ankara, Damascus and Baghdad (Waterbury 1990, p. 23). This committee has had rougher going than those formed with the USSR and Greece, for the actual volume of water in the Euphrates has been affected in a major way on at least two occasions.

Upon completion of the Keban Dam in 1974 Turkey began filling its reservoir at the same time that Syria began filling Lake Assad behind the Tabqa (Ath-Thawrah) Dam. This coincided with a major regional drought (see p. 49).

Again in January and February of 1990 Turkey reduced the flow of the Euphrates when it closed the spillways on the Ataturk Dam in order to complete construction on the river bed in front of the dam as well as to begin filling its reservoir. This event was clouded by the fact that Turgut Ozal, at that time Prime Minister, had hinted earlier that such action might be taken if incursions of PKK (Kurdish Socialist Workers Party) terrorists into Turkey from Syria and Iraq were to continue. This particular interpretation was not made explicit, however, and when the Syrians and the Iraqis sent their ministers to Ankara to protest the reduced flow of the river, the Turks replied that the matter was a technical one best worked out by the Joint Technical Committee. This was not accepted and the Iraqis in particular reasserted that the matter was political in nature. At that point the matter reached a stalemate and was not approached again until after the Kuwait War ended.

At the time of writing, it seems that discussions on sharing the waters of the Euphrates and Tigris Rivers are being pursued between the Syrians and the Turks, and a verbal agreement to guarantee that 500 cm crosses the border into Syria reconfirms an earlier accord on the same amount (see later, p. 70). It is also rumoured that the Karkamis Dam near the Syrian border in Turkey will not be built, and instead a small Syrian dam and reservoir will be allowed to impound the former's site. Of further note is the fact that during the Kuwait War when Premier Ozal was encouraged to shut off the river to 'punish' the Iraqis, he replied that the Turks would not use water as a weapon. This was in some part contradicted by the next Prime Minister, Suleyman Demirel, in July 1992, when he stated, 'Water resources are Turkey's and oil is theirs (Syria's and Iraq's). Since we don't tell them, "Look, we have a right to half your oil", they cannot lay claim to what is ours.' (*The*

Boston Globe, AP, 26 July 1992). It would seem, however, that such statements represent verbal jockeying for position and that serious talks will be (or are now being) carried out behind closed doors.

The South-east Anatolia Project (GAP) and Its Place in the Turkish Economy

The importance of the GAP to Turkey beyond hydropower and/or foreign exchange is obvious, given the underdevelopment of the south-east and the government's desire to stabilize the area politically through significantly raising the population's level of living. All of the above aims are summarized in the GAP *Master Plan*'s phrase that the region is to become an 'Agro-related Export Base'. It is in the context of both its positive and negative consequences that the GAP must be evaluated.

The importance of hydroelectric development through the creation of major dams and power plants provided the initial incentive for the GAP. The accompanying potential increase in irrigated cash crops and their sale abroad was thought of as a means of paying for such monumental structures. However, as the project attained reality, questions concerning the impact of GAP on the socioeconomic structure of both its region and the nation were inevitable. This led, in turn, to the creation of the aforementioned *Master Plan* through the cooperation of the Japanese firm Nippon Koei Co. Ltd. and the Turkish firm Yuksel Proje, A.S.

This consortium has defined the development objectives of the GAP as:

1. 'To raise the income levels in the GAP region by improving the economic structure in order to narrow the income disparity between the Region and other regions.'

2. 'To increase the productivity and employment opportunities in rural areas.'

3. 'To enhance the assimilative capacity of larger cities in the Region.'

4. 'To contribute to the national objective of sustained economic growth, export promotion, and social stability by efficient utilization of the Region's resources.'

These objectives are to be met through development strategies:

1. 'To develop and manage water and related land resources for irrigation, urban and industrial uses.'

2. 'To improve land use by managing cropping patterns and establishing better farming practices and farm management.'

3. To promote manufacturing industry with emphasis on agro-related ones and those based on indigenous resources.'

4. 'To provide better social services to meet the requirements of local people and to attract technical and administrative staff to stay in the region.' (*Master Plan*, Vol. 1, pp. 2, 3).

It should be noted that hydropower development is not specifically cited in these overarching objectives. Nevertheless, it continues to play an important part in development planning as is shown by three alternative scenarios suggested by the *Master Plan* in order to achieve the goals stated above. The first posits that all irrigation areas in the original plan be completed by 2005; the second, that power generation be maximized 'subject to the implementation of priority irrigation schemes'; the third, that 'only priority irrigation and hydropower schemes will be implemented by 2005' (*Master Plan*, Vol. 1, p. 9). In view of the difficulties surrounding the first two scenarios, the *Master Plan* recommends the third scenario. The reasoning behind that suggestion is summarized as follows:

GAP's original schedule called for the irrigation of one million hectares of land with Euphrates waters, and 625,000 ha with Tigris waters by the year 2002. To achieve this goal would require putting 100,000 ha into production in the Euphrates basin each year beginning in 1993, and another 60,000 ha per year in the basin of the Tigris. The annual ability of the General Directorate of Rural Affairs (GDRA) to develop farmland at present is thought to be between 1000 and 3000 ha (*Master Plan*, Vol. 2, p. 3.5). The slow pace of land redistribution, enmeshed as it is in a web of local and national politics, is largely responsible for this delay. At the same time, the Ministry of Agriculture, Fisheries, and Rural Affairs (MAFRA), responsible for the introduction of hands-on aspects of irrigated agriculture, is facing a Herculean task. Its facilities are limited and the inertia of the traditional farming systems is great.

Agricultural inertia may delay the realization of profits from agriculture for a number of reasons. The training of farm workers and the initiation of farm managers into the nuances of irrigated farming and the importance of applying exactly the right amount of water at the right time to the optimum mix of crops is a slow process. By the same token the fine tuning of agricultural choices to the best set of crops, matching conditions of soil, exposure, temperature and market demands, must be worked out in agricultural research stations, some of which are still fledgling institutions. Timing crops to the market and to the season of the year in order to catch demands at their peak also needs the development of additional skills. Finally, the proper packaging and marketing of crops, whether they are boxes of perishable fruit or bales of cotton, require even more expertise. For example, in the technically successful irrigation projects of the Menderes River in Western Turkey, I have seen piles of melons rotting by the roadside for want of customers (Kolars et al. 1983). All of these caveats are summed up by the *Master Plan* which states, 'The large discrepancy between the expected pace of irrigation development and the extremely limited capacity for on-farm works is a critical problem to be redressed' (Vol. 2, p. 3.5). My own personal experiences with village agriculture indicate that a twenty-year period of transition *after suitable facilities are in place* is more realistic then the proposed 2002 or 2005 finishing dates.

Table 8 shows the status of the GAP and Keban projects in 1992. If for simplicity's sake conjecture is limited to the Euphrates portion of GAP, we may anticipate that the Ataturk Dam and HEPP will be completely on line by about 1994. The Birecik Dam and HEPP will follow by about 1997. I personally feel that it is unlikely that a fourth dam—or fifth counting the Keban—the Krakamis, will be built. (See Kolars and Mitchell 1991, pp. 287–308). Thus, we must count the years until attainment not from the present, but in a stepwise fashion as the projects come on line one after another.

The second scenario, commitment to maximum hydroelectric production, is rejected by the *Master Plan* in view of the importance placed on the upgrading of economic and social conditions in the GAP region. Selling electric power at the expense of agriculture might benefit the nation but will fail to alleviate the poverty and its attendant unrest which are evident in the south-east.

The third scenario, the one recommended, essentially says 'take it easy', finish what has been begun, and let further investment and development be dictated by the pace of agricultural adjustments and social change. The practicality of this view cannot be denied. Nevertheless, such a course of action draws out the problems and decisions which river managers will have to face with increasing frequency in the years ahead. In order to consider what those may be, let us hypothesize a state of total development with every HEPP in place and every field capable of being irrigated, though not necessarily in production. What would the problems and choices facing management be under these conditions?

First, the condition of the rivers themselves, particularly the Euphrates, would present a challenge. A clearly recognized principle in the management of rivers with extreme seasonal and/or annual variance in

Table 8. Status of Construction and Development: GAP (1992)
(Including Production from Keban Dam)

	Euphrates Basin		Total Planned	Tigris Basin	
	In Product.	Under Construct.		In Product.	Under Construct.
Installed Power	3040 MW*	2852 MW	8716MW	—	402 MW
Energy Production	13254 GWh**	2450 GWh	33245 GWh	—	927 GWh
Irrigation ha	7330 ha	162051 ha	1623027 ha	—	52000 ha
No. of dams	3	3	22 + 1 dams	—	3
No. of HEPPs	2	2	19 + 1 HEPPs	—	3

Source: Kolars and Mitchell (1991, p. 37); EKA (1992).

* Keban installed power: 1240.

** Keban average annual est. production: 5900 kWh.

their flow (such as the Tigris and Euphrates) is that flood control as well as the storing of water for seasonal lows and periods of drought is best accomplished at the headwaters. (Water loss from evaporation is also minimized in this way—see Appendix, p. 92.) The Turks emphasize the advantages offered to Syria and Iraq by the large dams and reservoirs GAP provides. It was by such means in July 1987 that Turkey guaranteed a 500 cm (15.8 billion m³/yr) regular flow of the Euphrates across the border into Syria. The Turks also point out that this service has cost their downstream neighbours nothing, despite the high price of the Turkish dams. Nevertheless, the Syrians and the Iraqis remain unconvinced about the efficacy of the GAP and the goodwill of the Turks.

Their view can be understood if the case of full development is considered. Given such a situation, the GAP would have three significant negative impacts downstream: the *depletion* of downstream flow, the *detouring* of downstream flow, and the *pollution* of downstream flow.

Depletion of flow would come from evaporation from reservoir surfaces, necessary evapotranspiration for crop production, and system and on-farm inefficiencies resulting in unsalvageable water loss. With no irrigation, the Ataturk HEPP would have a firm discharge of 677 m³/s, firm annual energy production of 8190 GWh, and possible total annual production of 8705 GWh. Under full irrigation, firm discharge would be 375 m³/s, and firm annual production 4550 GWh (*Master Plan*, Vol. 2, table 5.4). It should be remembered as well, that some of the energy produced would be used to pump water to higher parts of the GAP fields. Similar values for discharge into Syria and total energy production for five dams (including the Keban and the Karkamis) are shown in Table 9 and Figure 1. The relationship is direct: the more agriculture, the less power. Thus, decisions will have to be made concerning the pay-off between agriculture and electricity.

The movement of water, energy, crops, manufactured goods, petroleum and foreign exchange, and the possible trade-offs among them, are shown in Figure 2. To begin with, water is lost through evaporation from reservoir surfaces. If reservoir water is used for irrigation there will be additional loss from evapotranspiration, though some return flow (RF) will occur in the system. Crops can be transferred to industry, or processed and sold directly for domestic consumption or foreign exchange. In every case electricity will be consumed to prepare the crops for sale. This electricity will come either from domestically produced hydropower or from thermal power (burning imported fuels). If water is used directly to produce electricity—with the minimum

Table 9. Estimated Future Relationships between Energy Production, Agricultural Water Use, and River Flow—the Euphrates River

Reservoir and Item	Without irrigation	With Urfa-Harran	Plus Mardin-Ceylanpinar	Full irrigation
Keban				
firm disch.	455 cm	465	465	465
firm energy	5200 GWh	5200	5200	5200
Karakaya				
firm disch.	562 cm	562	562	562
firm energy	6220 GWh	6220	6220	6220
Ataturk				
firm disch.	677 cm	627	546	375
firm energy	8190 GWh	7580	6610	4550
Birecik				
firm disch.	687 cm	637	554	347
firm energy	2330 GWh	2180	1940	1220
Karkamis				
firm disch.	688 cm	638	557	348
firm energy	583 GWh	536	473	284
Total				
firm energy	22523 GWh	21716	20443	17474
Secondary energy	2101 GWh	2136	2317	2597
Total energy	24624 GWh	23852	22760	20071

Source: *Master Plan*, Vol. 2, table 5.4.

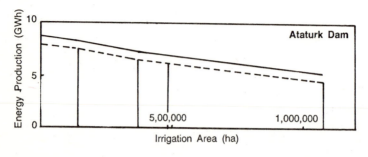

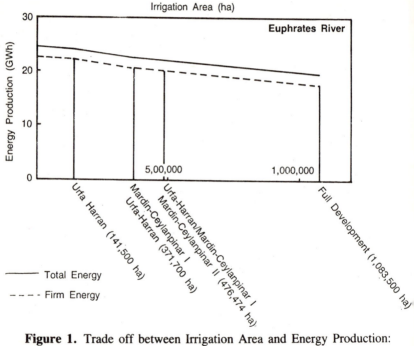

Figure 1. Trade off between Irrigation Area and Energy Production: The Euphrates River. (Source: *Master Plan*, Vol. 2, Table 5.4.)

amount of agriculture possible—additional energy can be exported to earn more foreign exchange, which in turn will offset the purchase of petroleum. At the same time, political and environmental dissonance will be minimized downstream.

If irrigated agriculture is emphasized, then electrical energy used to pump water to additional, higher fields, will add to the costs of agriculture. At the same time downstream dissonance is increased.

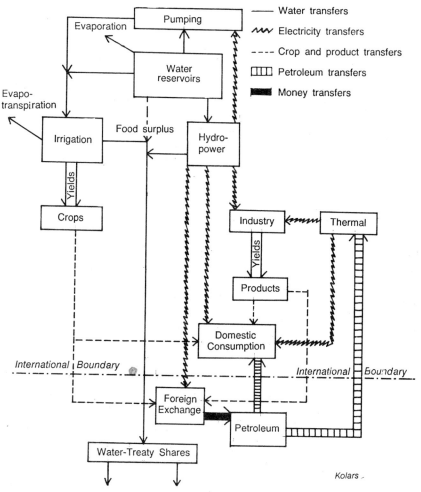

Figure 2. Trade offs among Irrigated Crops, Hydropower, Thermal Power, Foreign Exchange and Petroleum.

Thus, GAP managers will inevitably find themselves enmeshed in a nexus of competing demands. At stake will be not only the questions of more or fewer crops and what mix of crops to plant, but also questions of foreign exchange, petroleum imports, provision of cheap power for industry, and the assurance of sufficient and inexpensive food and industrial crops for home consumption or sale abroad.

Hydroelectric energy production, being a clean resource, could be

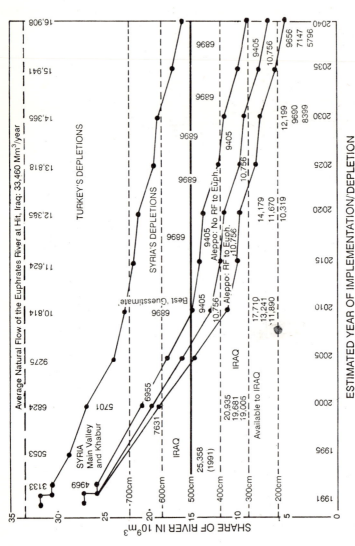

Figure 3. Projected sequential depletion of the Euphrates River, 1990–2040.

Note: Water Depletion of 10.814×10⁶ m³ labelled 'Best Guesstimate' is for Kolars and Mitchell's scenario for 666,697 ha irrigated and includes evaporation loss from reservoirs. This nearly coincides with the *Master Plan*'s estimated total water depletion of 10,429×10⁶ m³ for a net irrigated area of 931,411 ha (*Master Plan*, Table 5.3). The difference between these two figures represents a zone of uncertainty for river managers until further research or empirical evidence rectifies the disparity.

Source: Kolars and Mitchell 1991, tables 11.3–11.6, pp. 268–69.

favoured in spite of the *Master Plan*'s choice of the third scenario. However, internal political dissonance may become a major factor if electric production is increased at the expense of agriculture. Domestic pressure will be great to maintain irrigated crops at the expense of saleable power production. On the other hand, international, environmental dissonance resulting from reduced stream flow and lower reservoir levels in Syria and Iraq, as well as the possible pollution of such waters by dissolved solids, pesticides, herbicides and fertilizers, will place an opposite pressure on managers to favour power production over choices which will reverberate downstream.

The total water loss for irrigation associated with the Euphrates portion of GAP, as stated in the *Master Plan*, amounts to 10.429 billion m³ annually (Vol. 2, table 5.3, p. 5.27). I hope that this estimate is correct, although my own projections of annual water removal reach a possible 16.908 billion m³ (Kolars and Mitchell 1991, table 10.3, p. 208). Considering that on the average one cubic metre of water is necessary to meet evapotranspiration needs on each square metre of irrigated farmland—an assumption of which I am reasonably certain—and that there could be as much as one million ha of land irrigated after 2005, the *Master Plan* seems to meet evapotranspiration estimates without taking water loss (as described above) or evaporation from reservoir surfaces into account.[9] Thus, there is a possibility that as much as 60 per cent (16.9/28.2ths) of the flow into Syria might be pre-empted by agriculture. Even if my estimates run high, the discrepancy between the two figures is worrisome (Figure 3). Transboundary river flow under such conditions might average as little as 360 cm instead of the agreed upon 500 cm.

A second possible problem is the detouring of significant quantities of water from the main channel through the Urfa Tunnels to the Urfa-Harran and Mardin-Ceylanpinar irrigation projects (Figure 4). This means—according to my maximum projection—that 2.456 billion m³ might be subtracted from the main stream, i.e. transboundary border flow into Lake Assad, only to be returned to the main stream via the Khabur and Balikh tributaries downstream of Syria's main power plants and irrigation projects (Kolars 1922b). This flow would fulfil the agreed requirements—at least in part—but in a less than advantageous way for Syria. Lake Assad must remain full in order to feed the turbines at the Tabqa Dam, and to ensure the full capacity of the siphon or offtake

---

[9] The Turkish estimate of 10.4 billion m³ used annually approximates my own estimate of water necessary for the most likely total amount of irrigated land, approximately 700,000 ha, i.e. 10.8 billion m³ (Kolars and Mitchell 1991, Table 1.1.).

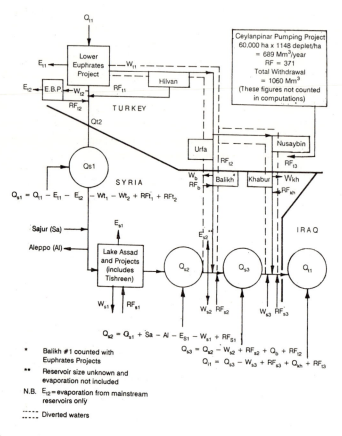

Figure 4. Sequential Water Budget of the Euphrates River *circa* 2010 + (*Source:* Kolars and Mitchell 1991, pp. 254–55.)

which provides the city of Aleppo with its water supply. Iraq, however, would not be affected by the detours discussed here.

Because the above-mentioned irrigation projects will be the first to come on line, this situation is one which could surface within the next few years, and may happen even though the initial amounts of water involved are much less than the possible aggregate after 2005. Again, Turkish managers may feel unexpected diplomatic pressure relating to this.

A variation on the obvious problem just described is that detoured water will be used in part to power a 50 MW generating plant at the Sanliurfa (south) end of the Urfa Tunnels. The estimated 124 GWh annual production will be assured only partially during the growing season when water will be directed to the Harran Plain by way of the Urfa HEPP facility. Additional, off-season flow may be continued in order to fill reservoirs intended to service additional fields in excess of the amount of water that can be delivered through the tunnels during the growing season. But when those reservoirs are full, will the water continue to flow through the tunnels to keep the Urfa HEPP operating? This flow will represent loss to Lake Assad, and as such will be another element in the water distribution and sharing equation to be solved by both Turkish and Syrian river management.

A third problem is the possible pollution of mainstream and tributary waters by the return flow from irrigated fields. It is certain that pollution of this kind will take place on the Euphrates and Tigris Rivers as more and more irrigation projects come on line. The *Master Plan* does not consider the international implications of this, and makes reference to drainage and salination in only one paragraph in its four volumes (Vol. 2, p. 5.6). Nevertheless, as river volume diminishes and water use increases, such problems will increase downstream. Syria may experience relatively little additional trouble regarding salination from Turkey, but its own soils are notoriously gypsiferous and saline, and their proper washing and cleansing could dump oppressive loads of dissolved solids on Iraqi fields. This problem is of less importance to Turkish river managers than either the detouring or the diminishing of river flow, but it must be kept in mind that reduced flow inevitably means greater concentrations of dissolved solids downstream.

Little has been publicly stated regarding the above matter at the time of writing. It is rumoured that the Syrians have asked the Turks to divert any extra water coming down the Balikh and Khabur Rivers eastward into topographical depressions in the eastern Syrian and north-western Iraqi deserts. Such a diversion should do little damage, either to fields in the Syrian Jezirah or to the evaporation pans in Iraq. Nevertheless, the engineering costs of such a solution to the problem would be considerable and the cooperation of the Iraqis would be absolutely essential. The loss of any water evaporated in this manner would also represent a significant loss to the entire system, particularly the portion of it in Iraq.

Impact on the Gulf

A further complication may wait at journey's end for the combined waters of the two rivers, the Shatt al-Arab. The Gulf into which the

Shatt empties is a shallow body of water with a circulation pattern requiring about three years. Its only opening is the Strait of Hormuz giving access to the Gulf of Oman. The Gulf has a rich and diverse marine life which supports the shrimp and pearling industries. Its coasts also support major desalination plants for Kuwait, Saudi Arabia, Bahrain, Qatar and the UAE. Its waters have a natural high salinity which is modified by the entry of fresh water from the Shatt, which also provides nutrients essential to the marine biota. If the flow of the Shatt is reduced as much as fifty per cent in the future, the water which reaches the Gulf will be considerably saltier than at present, and the ecology of the Gulf may be seriously affected. Higher salinities might also create additional problems for the desalination plants of the GCC. Although the impact on the Gulf is again of less immediate concern to Turkish management, its relation to the volumes of water maintained in Turkish reservoirs and flowing through Turkish fields can reflect upon Turkey's own problems. Thus, Turkey will find itself drawn with Syria and Iraq into river-associated relations with Gulf countries far from the rivers' headwaters (Figure 5) (Kolars 1992b).

Turkish Response to International Developments on the Euphrates

The Turks have been known for their diplomatic acumen from the founding of the Republic, through World War II, to the present time. President Ozal showed an awareness of the tensions growing over the impact of the GAP when in 1987 he suggested that a twin 'Peace Pipeline' might be built from the Seyhan and Ceyhan Rivers in southern Turkey to Sharjah in the UAE and to Jiddah on the Red Sea. Such pipelines would carry 3.5 thousand m³ per day (1.28 billion m³/year) in the west and 2.5 thousand m³ per day (0.9 billion m³/year) in the east at one-third the cost of comparable amounts of desalinized sea water. The two pipelines were at first estimated to cost $20 billion although more recent estimates give a much higher price (Brown and Root 1987). Nevertheless, the offer is still an important cooperative gesture which may possibly become part of the milieu in which river management will occur.

While such pipelines are technically feasible, the Arab states have viewed the offer with scepticism and no public avowals of interest. This stems from memories of Ottoman rule as well as practical fears that the pipelines could easily be cut by anyone, including other Arab states, 'upstream'. I have suggested that the latter problem can be overcome if water delivered by such pipelines is used to recharge exhausted aquifers

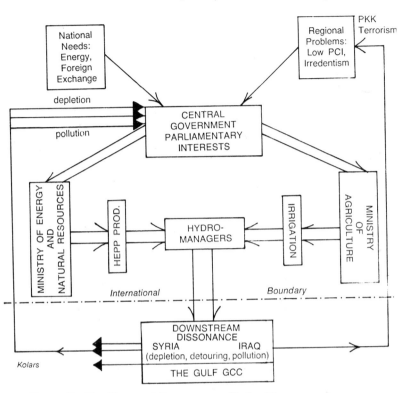

Figure 5. The Nexus of Pressures on Hydro Management.

rather than being introduced directly into domestic systems for immediate consumption. Thus, water already delivered would be 'money in the bank' for its recipients and could be used in times of severe drought. Moreover, if flow were cut, there would be only slight inconvenience to the recipients while they waited for it to be restored.

A second suggestion based on the original Peace Pipeline idea was made in 1991 by this author (Kolars 1991a, pp. 27–31) that a Mini-Peace Pipeline be built as far as Jordan, and that such a line use the waters of the Goksu or Manavgat Rivers west of the Seyhan and Ceyhan Rivers. (These alternative sources would be cleaner, more abundant, and more reliable then the original ones suggested.) This latter idea stimulated discussion among Israeli, Jordanian and Palestinian hydrologists faced with their own set of problems in the Jordan basin. Hillel Shuval (1992, pp. 139–40) has suggested an even shorter pipeline which would bring water from Turkey as far south as southern Syria. This line would have the advantage of being less expensive and directly involving only two countries. At the same time, it would free additional waters of the Yarmouk for use by Jordan and the West Bank. In this and other ways the zero-sum game in which finite and limited amounts of water are to be shared out by peace negotiators might be broken open.

Proposals such as these may seem unlikely when first considered, but given the Turkish water surplus noted at the beginning of this discussion, some similar project might actually be completed in the future. However, no Turkish politician could be expected to offer water to Turkey's neighbours while taps run dry in Ankara and Istanbul, or lights flicker dimly in Turkish homes. *The solution to this is that Turkey's rivers could be integrated into a single, comprehensive hydraulic system capable of smoothing intra-national/regional inequities and still leave something extra for international sharing.*

Piecemeal suggestions, however, can hardly be expected to be taken seriously by any of the nations concerned. On the other hand, there is an opportunity here for either the United Nations, or a consortium of interested nations, to help Turkey and its neighbours plan, build and operate a region-wide Middle-Eastern water network, a first step of which would be an integrated hydro-supply system for Turkey.

Syrian Developments and Reaction to Developments on the Euphrates

Syrian use of the Euphrates and its tributaries has changed significantly over time. The Sajur River which rises in Turkey is the first, and only, stream to join the main river in Syria from the right bank. Its reported flow (410 Mm³/year) was scheduled to a sustain a small Syrian irrigation project, but a recent Syrian commentary suggests that all of its waters are now consumed in Turkey (Mikhail 1992).[10] The first major Syrian use of the river is the Tishreen Dam with its relatively small holding reservoir (1300 Mm³). Continuing downstream, the Tabqa Dam forms Lake Assad with a storage capacity of 11,600 Mm³ and a surface area of 625 square kilometres.[11] The dam, which was designed and built with the help of Soviet engineers and follows Russian design, has its penstock openings located high on the upstream side. This makes it

[10] Turkish data give the annual flow of the Sajur as 138.6 Mm³, while other Syrian sources and the FAO give an average flow of 80.8 Mm³. These discrepancies are indicative of the need for standardized data (Kolars and Mitchell 1991, table 6.2).

[11] The materials given at this point are drawn from Kolars and Mitchell (1991).

necessary to maintain a high water level in the reservoir in order to use the dam's eight turbines for generating hydropower. In 1979 the dam provided 2.5 billion kWh, 60 per cent of Syria's electricity, but subsequent low water levels often have left the turbines idle. At the same time, an underground siphon brings water to the city of Aleppo from Lake Assad. Some 220,000 m^3/day (approximately 145 litres per capita), or about 80.3 Mm^3/year, were provided in 1980 in this manner for domestic purposes. Needless to say, with the rapid growth of Syria's urban populations, the city's dependence upon the Euphrates River becomes a matter of concern to Syrian planners. Downstream from the Tabqa Dam the next tributary to enter (from the left bank) is the Balikh which also rises in Turkey. Its flow is insignificant (190 Mm^3/year) and again is reported to be used completely within Turkey (Mikhail 1992). The final addition to the Euphrates is made by the Khabur River system which originates from a series of large springs just south of the border with Turkey. As described elsewhere, the water of these springs originates in Turkey and is subject to removal by pumping before reaching Syrian territory. There are reports that 'discharges of the Ras el-Ain Springs (recently) diminished (during the summer) by 6–10 m^3/sec. This is attributed to pumping from the aquifer in Turkey.' (Mikhail 1991, quoting G. Soumeh, 'Actual Use of the Al-Khabour River', Symposium on 'Water Resources Planning and Management', Aleppo University, 1987, pp. 253–63.)

It is difficult to assess the future of the Khabur, for on the one hand, detouring of Turkish irrigation water from Lake Ataturk via the Urfa Tunnels to the Khabur's headwaters might add several billion m^3 of additional return flow to the tributary. On the other hand, rerouting of such flow into desert evaporation basins (as discussed above) in combination with increased pumping of aquifers on the Turkish side of the border might reduce the flow of the stream. No further water enters the Euphrates beyond the confluence of the Khabur with the Euphrates near Deir ez-Zor.[12]

Understanding of the use of the Euphrates River and its tributaries in Syria for irrigation is obscured by lack of data and conflicting reports. Detailed analysis by Kolars and Mitchell (1991, pp. 274–82) indicates that much of the 640,000 ha originally scheduled for irrigation has had

---

[12] Reports indicate that Syria is now pumping quantities of water for irrigation from the Tigris River along the short stretch of that river which it shares internationally with Turkey. Little is known of the extent of this effort or its future impact on the flow of that stream.

to be abandoned because of gypsiferous soils. Two dams have been completed on the Khabur and a third is under construction near Hasakah which will facilitate use of land in the high Jezirah north of the Euphrates.[13] All in all, some 240,000 ha of public and private land were scheduled for irrigation or being irrigated (see p. 50) in the main valley in the late 1980s. Another 138,000 ha are planned or are being irrigated from the Khabur. A 'best guesstimate' by this author would be a final total 397,000 ha and a reduction of river flow by 4.7 billion m^3 annually. This does not include, however, a possible 200,000 ha near Aleppo which may be irrigated with water from Lake Assad. Such removals could reduce the flow in the Euphrates by 2.5 billion Mm^3/year if return flow from the fields is pumped back to the main stream, or by some 3.9 billion m^3/year if the water is lost locally to evaporation.

Syrian river managers are thus faced with two major types of problems. They must respond to Turkish manipulation of the river upstream and they must balance their own priorities internally. The Turks have agreed to allow at least 500 cm (15.8 billion m^3/year) across the border into Syria, although there are periodic rumours of Syrian demands for 700 cm (22 billion m^3/year). Just how depletions or additional flow from Turkey on the Khabur will be accounted is uncertain. Meanwhile, Lake Assad must be kept full if its hydroelectric potential is to be realized. This, in turn, conflicts with increased demands for domestic water for Aleppo as well as for the various irrigation projects mentioned above. While only a relatively small volume of water will be needed for Aleppo, its purity and its assured flow will be of paramount importance. Thus Syria, to date, is essentially reacting to events on the river. How Syria's future role will be resolved vis-à-vis sharing the Euphrates, and the possibility of a Peace or Mini-Peace Pipeline across Syrian territory remains to be seen.

Iraq's Use of the Euphrates River

In both ancient and modern times, the management of the Euphrates River had its beginnings in Iraq. A discussion of ancient water management is superfluous for this paper, except that those early beginnings partially establish Iraqi claims to use of the river. The long-term and extensive use of river water in Mesopotamia also positions Iraq vis-à-vis the model presented in the first part of this discussion.

River management in modern times began with the report of British

[13] A fourth dam is apparently planned for the Jezirah, but its site and characteristics are not available at the time of writing.

hydrological engineer William Wilcox to the Ottoman Empire in 1911. His suggestions included 'the al-Hindiya Barrage on the Euphrates [completed in 1913], the Kut Barrage on the Tigris, the Habbaniya projects, the Tharthar project, the Naharavan irrigation project, [the] Bekhme Dam, and the Mosul Dam' (Naff 1991, p. 4). Under the British Mandate (1917–32), which began the collection of pertinent data, a Department of Irrigation was established in 1918.[14] This is not to say that the river was unexploited prior to the British occupation. Traditional canals and weirs had continued in use over the centuries, but the system was unplanned, uncoordinated, and had fallen into serious disrepair.

In the years that followed, the Kingdom of Iraq created a Board of Development, the Ministry of Development, and the Ministry of Agrarian Reform. An intensive programme of planning followed with the help of foreign firms; however, comprehensive integration of the programme was disrupted by the revolution of 1958. It has continued to be kept off balance by the subsequent actions of the new Iraqi government.

Agriculture received special attention following the nationalization of the oil industry in 1972 with the establishment of the Higher Agricultural Council (attached to the presidency), the Land Reclamation Organization and the Ministry of Irrigation. A comprehensive master plan, 'General Scheme for Planning Water and Land Resources of Iraq' (unavailable to this author), was developed with the help of the Soviet Union from 1970 to 1984. The Master Plan is reported to cover every aspect of land and water use in the country and to project such development and planning up to the year 2000. However, this emphasis was short-lived, and in 1979 Saddam Hussein abolished the Ministry of Agrarian Reform and combined the Ministry of Agriculture (established in 1970) and the Ministry of Irrigation with a 30 per cent reduction in staff. Since that time, the Iran-Iraq War and the invasion of Kuwait have diverted attention from agriculture and hydrologic development.

The existing and planned development and use of the Euphrates and Tigris Rivers in Iraq is shown on Map 3. The Haditha (Qadisiya) Dam, completed in 1987, with a storage capacity of 7 billion m^3 is farthest upstream, on the Iraqi portion of the Euphrates. This dam is meant to generate hydropower and to provide water for as yet uncompleted irrigation projects. The Baghdadi Dam 40 km south of the Haditha Dam will regulate flow from the latter structure. Work on a preliminary coffer dam was begun in 1990. Downstream, the Ramadi Barrage is

[14] Discussion of Iraq's water situation is drawn in large part from Naff (1991) and to a lesser degree from U.S. Corps of Army Engineers (1991).

used to divert water into the Habbaniya Reservoir whence it can either be returned to the main stream via the Dibban Canal or permanently drained off to the Abu Dibbis Reservoir. Farther downstream, the Hindiya Barrage is used to divert water for irrigation and to maintain a necessary head of water for gravity flow to adjacent fields. The Fallouja Dam (diversion) and the Hammourabi Dams (regulation)[15] complete the manipulation of the Euphrates in Iraq.

Highest on the Tigris River in Iraq is the Mosul (Saddam) Dam used for hydropower, irrigation and flood control. Next downstream, the Badush Dam, in the design and planning stage, is a safety resort in the event of damage to the Mosul Dam. Work was begun on the Badush in 1988 but it is not yet completed. The multipurpose Fatha Dam below the confluence of the Lesser Zab and the main stream is in the planning stage. The Samarra Barrage (1950s) is used to divert flood control water into the Tharthar depression, while the al-Kut Barrage (1939) farthest downstream is used for irrigation and some flood control.

Developments on the eastern tributaries of the Tigris must also be considered. The Bekhme Dam (storage capacity 12 billion m³) on the Greater Zab has already been mentioned. Work started on this dam in 1989 but has been delayed indefinitely by the Kuwait War. Also on the Greater Zab and its tributaries are the Khazir-Gomel and Mandawa Dams—in the design and planning stages—which are intended for irrigation and regulation of flow. The Dokan Dam on the Lesser Zab (1959) is primarily meant for flood control. The Dibbis Dam on the same stream is used for irrigation. Three dams on the Adhaim River are at the design stage, while the Darbandikhan Dam (1961) and the Hamrin Dam (1987) on the Diyala, and the Diyala Weir (1928) are for irrigation and flood control.

The Main Outfall Drain (the Saddam River or the Third River), 500 km in length, with an average depth of 4 metres and a width of 180 metres, is also of note. This impressive canal is intended to remove excess drainage water from the area between the twin rivers south of Baghdad and to discharge it into the Gulf near the Fao Peninsula after tranferring it by siphon across the Euphrates River near Nasiriyah. Ninety per cent finished in 1991, unofficial information indicates that it has recently been completed.

Persistent rumours assign a more sinister role to the dams in northern Iraq and to the Main Outfall Drain. In the former case, the reservoirs are

---

[15] Exact locations unknown to this author, but below the Hindiya Barrage.

said to be intended to fragment the territory of the Kurds, while in the south, the Shiites are the targets. Certainly, water barriers have been used by the Iraqis before. The creation of 'Fish Lake' on the south-eastern frontier with Iran during the Iran-Iraq War is an example of this. In the former cases, however, the peaceful utility of the reservoirs far outweighs any more bellicose motivation.

Estimates of the actual amount of land irrigated with Euphrates' waters in Iraq vary from author to author. Chalabi and Majzoub(1992, table 2) indicate that 1.2 million hectares were thus farmed in the 1980s in Iraq. Their projection of irrigated land using Euphrates' water after the year 2000 reaches 1.8 million ha. (Their figures for Turkey are 150,000 ha and 1.25 million ha respectively, and 250,000 ha and 795,000 ha for Syria.) Any such estimates must be modified in terms of delivery system and farm system efficiencies and their improvement or continued deterioration.[16]

The complexity, secrecy and confusion surrounding river water use in Iraq makes analysis of the Iraqi response to development of the Euphrates difficult. Interpretation is further clouded by the disagreement among numerous estimates both of present use and future need, not only in Iraq but also in Syria and Turkey. Tables 10 and 11 show one set of estimates for the natural flow and future use of the two rivers. The full-use scenario for the Euphrates may seem unrealistically high, yet one should not forget that the Colorado River now only reaches the Gulf of California in exceptional high-flood years. Nor is it possible to predict with certainty how much development will be achieved, and therefore how much water each country will need, in the years ahead. The one evident and incontrovertible conclusion is that there will not be enough water in the Euphrates to satisfy every demand, no matter how modest is the scenario chosen.

One possibility which might help the Iraqis maintain control of their hydrologic future is the water remaining in the Tigris River. Ignoring for the moment the political difficulties surrounding the question of Kurdish rights and sovereignty and the difficulty of development work in what amounts to a war zone at the present time, a canal might be built from the Mosul reservoir (or from a smaller retaining or diversion facility farther upstream) in order to bring a supplemental supply of water to the Euphrates River. Such a canal could run almost straight south following the 500-metre contour to the Euphrates below the

---

[16] Kolars (1991a) discusses other estimates of irrigated areas in Iraq. The amounts are of the same magnitude as discussed herein.

Table 10. Sources and Uses of the Euphrates River (Mm^3/year)

Natural Flow	Observed at Hit, Iraq	29,800
	Removed in Turkey (pre-GAP)	820
	Removed in Syria (pre-Tabqa)	2,100
	Natural flow at Hit	32,720
Pre-Keban Dam (< 1974)	Flow in Turkey	30,670
	Removed in Turkey	− 820
	Entering Syria	29,850
	Added in Syria	+2,050
	Removed in Syria	−2,100
	Entering Iraq	29,800
	Added in Iraq	—
	Iraqi irrigation	−17,000
	Iraqi return flow (est.)	+ 4,000
	To Shatt Al-Arab	16,800
Full Use Scenario (*circa* 2040)	Flow in Turkey	30,670
	Removed in Turkey	−21,600
	Entering Syria	9,070
	Removed in Syria	−11,995
	RF & Tributaries (Turkey/Syria)	+9,484
	Entering Iraq	6,559
	(Removed in Iraq)	(− 17,000)
	(RF in Iraq)	(+ 4,000)
	(Deficit to the Shatt Al-Arab)	(− 6,441)

Source: Kolars (1992a)

Table 11. Sources and Uses of the Tigris River
(Mm³/year)

	Pre-project	After 2000	Natural Flow
Flow from Turkey	18,500	18,500	18,500
Removed in Turkey	—	− 6,700	
Entering Iraq	18,500	11,800	
Inflows to Mosul	2,000	2,000	2,000
Greater Zab	13,100	13,100	13,100
Lesser Zab	7,200	7,200	7,200
Other	2,200	2,200	2,200
Subtotal	43,000	36,300	43,000
Reservoir Evaporation	—	− 4,000	
Irrigation (to Fatha)	− 4,200	− 4,200	
Return Flow	+ 1,100	+ 1,100	
Adhaim River	+ 800	+ 800	800
Irrigation (to Baghdad)	− 14,000	− 14,00	
Return Flow	+ 3,600	+ 3,600	
Domestic Use	− 1,200	− 1,900	
Diyala River	+ 5,400	+ 5,400	5,400
Irrigation	− 5,100	− 5,100	
Return Flow	+ 1,300	+ 1,600	
Subtotal	30,700	19,200	49,200
Reservoir Evaporation	—	900	
Irrigation to Tokut	− 8,600	− 8,600	
Return Flow	+ 2,200	(2,200 to Outfall Drain)	
Total to Shatt Al-Arab	24,300	9,700	49,200

Source: Kolars (1992a)

Haditha Dam. This, in combination with water stored in reservoirs on the eastern tributaries of the Tigris, might alleviate Iraq's predicted water problems. The expenditure on such ventures should be considered as an international, regional item to be shared by all the riparians. Such an idea raises the possibilities of potential basin-wide/regional cooperation.

Potential Basin-wide/Regional Cooperation

Before examining the subject of such cooperation, note must be made of a difference in point of view among the riparians. Turkey claims that the Euphrates and the Tigris Rivers are 'transboundary' rather than 'international' rivers, and therefore the Turks have the right to control development of these rivers in an 'equitable, reasonable and optimal manner' (Chalabi and Majzoub 1992, p. 19). Three stages would be necessary: (1) to make a detailed inventory of hydrologic information, to normalize all measurements and techniques thereof, and to openly exchange all necessary data; (2) to conduct an inventory of all arable lands in the basin in order to establish the optimal crops for each area as well as the amount of water needed for the successful cultivation of such crops; and (3) to combine the above data in order to ascertain the optimal division and use of land and water within the entire basin, including water lost through evaporation from reservoirs and also possible water transfers from the Tigris to the Euphrates. It should be noted that, as yet, no mention has been made of some manner of environmental advocacy for the rivers themselves nor of the impact on the Gulf of the scheduled and anticipated developments.

The Arab nations strongly object to such an overarching attitude on the part of the Turks (Chalabi and Majzoub 1992). Their major argument is that international precedent, if not enforced law, insists that the management and sharing of rivers be equally in the hands of all the riparians involved. It is not the purpose of this discussion to attempt to untangle this political Gordian Knot. It seems more appropriate to close with a confessedly optimistic consideration of what could transpire if such antithetic attitudes were to be resolved.

A strong case for regional cooperation regarding water has been made by Shawki Barghouti of the World Bank. He emphasizes that the proper management of water resources in the Middle East is beyond the ability of any single nation, and calls for a minimum hydrological planning unit 'such as an entire drainage basin'. Three elements would be critical for basin-wide cooperation: a free exchange of data and information on the water sector, a genuine regional water management establishment, and the application of modern water technology and engineering (Barghouti 1992). Such a suggestion raises complex questions of international cooperation and law (Solanes 1992; Moore 1922, pp. 16–77).

The acquisition, verification and analysis of data underlie all water management. Data regarding stream flow, precipitation, evapotranspiration, water removals, return flow, salinity and a host of other variables are notoriously scarce, incomplete and open to question everywhere in the Middle East. Nations have, until now, viewed data as knowledge, and by extension, data as power. Obtaining good water-related data has been likened to counting sheep in the desert. If a fodder supplement is being offered there are plenty of sheep. If a head tax is proposed, there are very few sheep. Successful negotiations between parties contending for limited amounts of water can only succeed in the long term if agreements are based on an accurate picture of what water is available (Kolars 1991b).

Although Turkey has demonstrated its own reservations about sharing data, it has also shown its goodwill through negotiations such as those described above with Iraq and Greece. Turkey is now on the verge of acquiring technology which will put it in the forefront of data acquirers and sharers in the Middle East and the new nations of central Asia. The Earth Observation Satellite Company (EOSAT) is at present negotiating with the Government of Turkey to provide it with an archive of past satellite imagery, as a first phase of a continuing programme. The archival material would provide recent historical coverage of natural conditions in Turkey, the adjacent Middle East, and the Central Asian Republics. Turkish efforts could be easily coordinated—technically, if not politically—with similar work using Geographical Information Systems (GIS) conducted by the Government of Qatar.

The Euphrates-Tigris River basin, dependent as it is upon conditions at its mountain headwaters for both replenishment and long-term storage, presents another unique opportunity for Turkish cooperation in addition to the sharing of data. Reservoirs can not only be used to store water in the political unit where they are located, but may be viewed as part of a *water bank* wherein 'reservoir space is allocated to different contracting entities and is based on criteria developed by water experts from the participating states' (Barghouti 1992, p. 10). This technique has already proved successful in California. Contracts giving fixed amounts of reservoir storage are let to specific users. In years with an abundance of water contracted water remains in the reservoir to the space holder's credit. In drought years water can be released to downstream contractors according to existing agreements. Water can also be sold or traded to other users 'as long as existing contractors will not suffer shortages as a result of new use' (Barghouti 1992, p. 10).

Such an arrangement may seem unlikely, given the suspicion evident among the three riparians involved on the Euphrates, but a nexus of related agreements would help to guarantee all of them being equally honoured. For example, water destined for southern Syria (as per Kolars' Mini, or Shuval's Mini-mini pipelines) could guarantee the passage of Iraq's 'water bank balance' downstream through Syria to its contractor. Petroleum transfers to Turkey could insure the flow of water to Syria. A further refinement of such agreements could be the equating of certain amounts of hydroelectricity or crops produced in Turkey or

Syria in exchange for equivalent quantities of water diverted and lost through upstream agricultural evapotranspiration. It should also be considered at this point whether evaporation from upstream reservoirs (not an insignificant quantity) should be deducted on a proportionate basis from the water bank accounts of downstream partners as their fair share of operating expenses.

The possibility of such an intra-regional system of cooperation has advanced a step forward with the Turkish Minister of Energy and Natural Resources, Ersin Faralyali's announcement that Turkey, Syria, Egypt and Jordan have signed an agreement (October 1992) for the exchange of surplus supplies of electrical energy. Feasibility studies for the international project have been completed, and 'agreement has been reached on general trade, installations and interconnections' (*Newspot*, 22 October 1992, p. 5).

The third area of cooperation suggested by Barghouti—modern technology and engineering skills—finds application in the Turkish case. GAP, the articulated end result of Turkish development planning, is proof that Turkey has come of age technologically. Elsewhere, the construction of the Bekhme Dam on the Greater Zab was about to be begun by a Turkish company, when it was interrupted by the invasion of Kuwait (Naff 1991, p. 38). Other examples of the export of Turkish engineering skills include land-levelling for irrigation at numerous locations throughout the Middle East and a variety of contracts given to Turkish firms by the Libyan government. The impact of Turkey's acquired engineering skills is demonstrated by the recent signing of a contract by the Turkish Bureau of Associated Engineers (BMB) to manage four oil fields and to construct a major power plant in Kazakhistan (*Turkey Today* 1992, p. 5). Why not hydroelectric management skills as well?

Conclusion

If a modern Jules Verne were to cast his imagination into the Middle East and its hydraulic future, he might foresee a time when hydrologists, agronomists, hydroelectric engineers and diplomats would meet at a centre which would incorporate among its many facilities satellite receiving stations transmitting not only immediate satellite images of weather conditions but also the readings of scores of remote stations monitoring temperature, snowfall, stream flow, reservoir levels (*ergo*, holdings) and soil moisture.

These, in turn, would be synthesized with archival records by GIS technology into a comprehensive picture of water availability, probabilities, and water needs throughout the international region. Diplomatic brokers might then bargain and exchange supplies of electric power, stored water, petroleum, or available food and industrial crops using techniques similar to those suggested by James W. Moore in his study of pragmatic and numerical approaches to water-sharing regimes in Israel and the Occupied Territories (Moore 1992), whereby comparative values can be placed on the dissimilar elements described above.

It is possible to imagine Turkish water flowing through pipelines to recharge aquifers in Saudi Arabia, or perhaps mingle with desalinized sea water from plants located on the coast of Oman, in turn to sustain the people of Gaza or to recharge aquifers threatened by salt water intrusion along the shores of the UAE. Or again, Turkey might cash in hydro-credits or excess electric power for petroleum from Iraq or natural gas from Syria. The Central Asian Republics might even find chances for partnership in such a far-reaching cooperative network.

Whatever the scenario, Turkey's pivotal geographic location plus its surplus supplies of water make it a central actor in the rational management of Middle Eastern resources of water, oil, land and food, and energy. Sustainable development and the peace attendant upon it can only be achieved through ceaseless efforts by the three riparians which share the Euphrates-Tigris basin. It is necessary to forget past history and cultural differences and, Arab and Turk alike, strive for rational management of the entire system in order to make it a keystone in an overarching regional peace.

APPENDIX

Surface Areas and Volumes of Some Middle-Eastern Reservoirs

Country	Dam/Reservoir	Volume (V) (1×10^6 m³)	Area (A) (1×10^6 m²)	Ratio: V/A
Turkey	Keban	30600	675	44.4
	Karakaya	9580	298	32.1
	Ataturk	48700	817	59.6
	Birecik*	1220	56.25	21.7
	Karkamis*	200	28.4	7.0
Syria	Tishreen	1300	70	18.6
	Tabqa (Ath Thawrah)	11700	628	18.6
	Ba'ath	90	2.7	33.3
Iraq	Haditha (Qadisiya)	10000	?	—
	Fallouja	3600	?	—
Egypt	Lake Nasser	78500	3500	22.4

Sources: Kolars and Mitchell (1991) and U.S. Army Corps of Engineers (1991) (Computations by author)
* to be built.

The ratio in the table indicates the number of cubic metres of water, as it were, beneath each square metre of reservoir area. The larger the number, the more efficient the storage vis-à-vis evaporation losses. Mountain (i.e. headwater) locations provide the best and deepest reservoir sites. In the case of the Euphrates reservoirs, it should be noted that the farther downstream the reservoir in question is located, the higher will be the average annual ambient air temperature, resulting in greater evaporation losses per square metre of surface. This constitutes a multiplier effect when considering the best (or worst) places to store water.

BIBLIOGRAPHY

Aydinelli, Sevket. 'Turkiyenin Energi Ekonomisi ve Elektriklendirilmesi' (Elektrik Isleri Etut Idaresi Nesriyati, Ankara: 1940), paper, 175 pp. with fold-out end-maps.

Barghouti, Shawki. 'Water Resources in the Middle East—an Agenda for Regional Cooperation', paper presented to the Council on Foreign Relations (New York City: 17 June 1992), 12 pp.

Bilen, Ozden and Savas Uskay. 'Background Report on Comprehensive Water Resources Management Policies—An Analysis of the Turkish Experience', paper presented at the World Bank International Workshop on Comprehensive Water Resources Management Policies (Washington, D.C.: 24–28 June 1991), 110 pp.

Biswas, Asit K. 'Management of International Waters—Problems and Perspectives' (International Society for Ecological Modelling, Oxford: 1993), 36 pp. (Revised version included in this volume.)

The Boston Globe. 'Turkey Opens Big Dam on Euphrates River', (AP: 26 July 1992).

Brown and Root, Inc. 'Source to Consumer' (Houston: 1987) 10 pp.

Chalabi, Hassan and Tarek Majzoub. 'La Turquie, les eaux de l'Euphrate et le doit international', (Water in the Middle East—Legal, Political and Commercial Implications) (SOAS, London: 19–29 November 1992), 32 pp.

Devlet Su Isleri. *Technical Bulletin with Maps—1983* (DSI Printing House, Ankara: 1/19/1984), 122 pp. including fold-out maps.

——. *Turkiye'de Barajlar ve Hidrolelectrik Santrallar—1984* (Isletme Mudurlugu Matbaasi, Ankara: 1984), 24 pp. plus end-map.

EKA International. 'Heavy Construction Activities Continue in GAP Region', Special Issue on GAP (1992), pages unnumbered.

Evans, T. E. 'History of Nile Flows', *The Nile—Resource Evaluation, Resource Management, Hydropolitics and Legal Issues* (SOAS/RGS, London: 1990), pp. 5–28.

Faralyali, Ersin. 'The Ataturk Hydroelectric Power Plant and Turkey's Electric Energy Sector', *EKA International*, Special Issue on GAP (1992), pages unnumbered.

Hinnebusch, Raymond A. *Peasant and Bureaucracy in Ba'thist Syria* (Westview Press, Boulder: 1989), 325 pp.

Kolars, John. 'The Hydro-Imperative of Turkey's Search for Energy', *The Middle East Journal* (Vol. 40, No. 1, Winter 1986), pp. 53–67.

——. 'A Model of Middle Eastern River Basin Development at the National Level', paper presented to the Middle East Studies Association Annual Meetings, San Antonio, Texas (1990).

——. 'The Future of the Euphrates River', Prepared for the World Bank Conference: International Workshop on Comprehensive Water Resources Management Policy (Washington, D.C.: 24–28 June 1991a), 33 pp.

——. 'The Role of Geographic Information Systems (GIS) Technology in the Future Management of Middle Eastern Rivers', paper presented at the 1991 MESA Annual Meeting (Washington, D.C.: 23–26 November 1991b), 9 pp.

——. 'Water Resources of the Middle East', *Canadian Journal of Development Studies* (Special Issue, 1992a), pp. 103–119.

——. 'Fine Tuning the Euphrates-Tigris System', paper presented to the Council of Foreign Relations (New York City: 17 June 1992b), pp. 18.

——. 'The Challenge of Managing Mega-hydro Projects: the Ataturk Dam and the Southeast Anatolia Development Project', *Papers Presented to the Financial Times Conference: 'World Electricity'* (*The Financial Times*, London: 9/10 November 1992c).

Kolars, John and W. A. Mitchell, *The Euphrates River and the Southeast Anatolia Development Project* (Southern Illinois University Press, Carbondale: 1991), 317 pp.

Kolars, John, J. Casstevens and J. Wilson. *On-farm Water Management in Aegean Turkey—1968–1974*, AID Project Impact Evaluation Report No. 50 (U.S. Agency for International Development, Washington, D.C.: 1983), 52 pp.

Master Plan, see below, Nippon Koei Co. Ltd. . . .

Mikhail, Wakil. 'Water Resources Development Plans in Syria: role of the Euphrates Basin', (draft ms. 1992), 24 pp.

Moore, James W. 'Water Sharing Regimes in Israel and the Occupied Territories—a Technical Analysis', Project Report 609 (Department of National Defense, Operational Research and Analysis Establishment: Ottawa: August 1992), 46 pp.

Naff, Thomas. 'Water Issues in Iraq', prepared for the Associates for Middle East Research (AMER, Philadelphia: revised August 1991), 66 pp.

Naff, Thomas and Ruth C. Matson, ed. *Water in the Middle-East—Conflict or Cooperation?* (Westview Press, Boulder: 1984), 236 pp.

Newspot. 'Energy Agreement' (22 October 1992), p. 5.

Nippon Koei Co. Ltd. and Yuksel Proje A.S., The Southeast Anatolia Project Master Plan Study, *Final Master Plan Report*, Vol. 1: *Executive Summary*; Vol. 2: *Master Plan* (Ankara: April 1989).

Sanlaville, P. and J. Metral. 'Water, Land, and Man in the Syrian Countryside' (in French), *Revue de Geographie de Lyon* (Number 51: 1979), pp. 220–40.

Shuval, Hillel I. 'Approaches to Resolving Water Conflicts between Israel and Her Neighbors—a Regional Water-for-Peace Plan', *Water International*, 17 (1992), pp. 133–43.

Solanes, Miguel, 'Legal and Institutional Aspects of Basin Development', *Water International*, 17 (1992), pp. 116–23.

Turkey, Government of, State Hydraulic Department (DSI). *1983 Statistical Bulletin with Maps* (Ankara: 1984), 114 pp.

———. State Institute of Statistics, *Statistical Pocketbook of Turkey—1990* (State Institute of Statistics Printing Division, Ankara: January 1991), 318 pp.

———. *Statistical Yearbook of Turkey—1989* (SIS Printing Division, Ankara: 1990), 447 pp.

Turkey Today. 'Major Energy Contract for Turkey', Number 134 (July/August 1992), p. 5.

United States Corps of Army Engineers. *Water in the Sand—A Survey of Middle East Water Issues* (Washington, D.C.: 1991), pp. iv and 155.

Waterbury, John. *The Hydropolitics of the Nile Valley* (Syracuse University Press, Syracuse: 1979), 301 pp.

———. 'Dynamics of Basin-wide Cooperation in the Utilization of the Euphrates', paper prepared for conference on 'The Economic Development of Syria: Problems, Progress, and Prospects', Damascus (6–7 January 1990).

Weatherford, Gary D. and F. Lee Brown, eds. *New Courses for the Colorado—Major Issues for the Next Century* (University of New Mexico Press, Albuquerque: 1983), 253 pp.

4 / Prospects for Technical Cooperation in the Euphrates-Tigris Basin

ÖZDEN BILEN

1. INTRODUCTION

Over the last decade, numerous reviews and studies have appeared in books and journals addressing the Middle East water issues. These have generally been diluted with multiple political arguments. A very sensitive and misused commodity by nature, water has often been manipulated by human beings driven by different aspirations and political ambitions. Several divergent views and political arguments have emerged concerning the use of transboundary rivers. It is not, however, the intention of this paper to go into the merits and demerits of these distinct approaches; rather it attempts to indicate some technical solutions which could help determine the supply and demand balance of the water required by the riparian countries in the Euphrates-Tigris basin.

Public perception of water issues is generally influenced by politics, and as such is highly ambiguous. In this context, a scientific and technological approach is required in order to comprehend the issues clearly and objectively.

2. WATER RESOURCES MANAGEMENT AND ALLOCATION ISSUES IN THE EUPHRATES-TIGRIS BASIN

It would be useful to define water resources management as 'the art of matching the supply of water with the demand while controlling the quality'. Throughout this process, we confront two complementary policies:

(i) *Supply-augmenting policies*, which include making use of storage facilities, water transfer among rivers and non-conventional water supply methods.

(ii) *Demand-management policies*, which include making more efficient use of existing supplies through structural, operational and economic means. Demand management could reduce the scale of supply-augmenting projects or even remove the need for them.

These two policies are treated inseparably while setting up project alternatives and searching for optimum solutions. However, this paper is mainly confined to the engineering aspects of water resources management.

2.1 Optimal Run-Off Regulation in the Basin

Large annual and seasonal variations observed in the run-off of most large basins make it necessary for water resources managements to store adequate water in the upper catchments in order to allow regulated flows throughout the year and over the years.

The seasonal and annual flows of the Euphrates and the Tigris rivers have extremely high variance. At the Birecik gauging station on the Euphrates near the Syrian border, the average annual flow is 30.5 km^3. Two distinct dry cycles were recorded over the 1937–90 period. The first was in 1958–62, 1961 being the year with the most severe shortfall when the annual flow was as low as 14.9 km^3 which accounts for just 49 per cent of the long-term average. The second dry cycle started in 1970 and ended in 1975. The lowest flow was in 1973 with an annual flow of 18.8 km^3, representing 62 per cent of the average. On the other hand, the recorded peaks of annual flow were 56.4 km^3 and 57.7 km^3 in 1969 and 1988, respectively. These represent 185 per cent and 189 per cent of the long-term average. The flow rate of the Euphrates also has significant seasonal variations. In an average year, the highest flow is generally observed in April or May and the lowest in September. The fact that the monthly flow of the Euphrates fluctuates between 530 per cent and 16 per cent of the monthly long-term average is sufficient evidence of the high seasonal fluctuations (DSI 1992).

The historical records of the annual and monthly run-offs occurring on the Euphrates river at Belkisköy (Birecik) in the 1937–80 period are illustrated in Figures 1 and 2, respectively.

Similar high seasonal and annual fluctuations are also observed in the Tigris river. According to the discharge records at Cizre gauging station on the Tigris near Turkey's border with Syria, the annual average flow was 16.8 km^3 over the 1969–90 period. The Tigris' annual flow variations are similar to those of the Euphrates. The 1970–75 period experienced a drastic decline in the flow rate, the lowest being in 1973 at 9.6 km^3, corresponding to 58 per cent of the average. On the other hand, 1969 was a peak year with 34.3 km^3 measured at Cizre station (204 per cent of the annual average) (DSI 1992).

The historical records of annual and monthly run-offs occurring on the Tigris river in the 1946–82 period at Cizre are also illustrated in Figures 1 and 2, respectively.

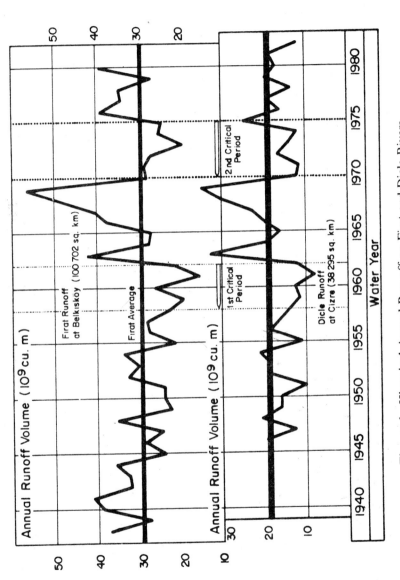

Figure 1. Historical Annual Run-off on Firat and Dicle Rivers.
Source: Bagis (1989, pp. 43–44).

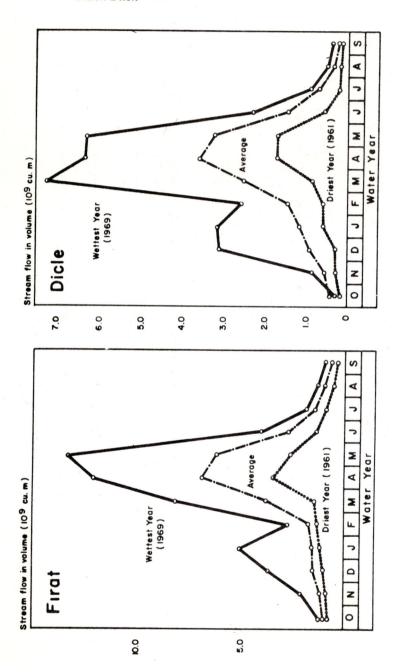

Figure 2. Monthly Run-off on Firat and Dicle Rivers.
Source: Bagis (1989, pp. 43–44).

Because of the extremely high seasonal and annual flow fluctuations in the Euphrates and the Tigris rivers, storage facilities are a key concern in the problem of water resources management for the riparian countries in the Euphrates-Tigris basin. However, the Euphrates, along its entire course in downstream countries, does not provide ideal sites for the creation of large dams and associated reservoirs (Figure 3). The largest dam in Syria (Tabqa) has only 9 km^3 active storage capacity which accounts for only 30 per cent of the natural flow of the Euphrates. Main storage facilities, existing or planned, on the Euphrates river in Turkey are Keban, Karakaya, Atatürk, Birecik and Karkamiş dams, of which Keban and Karakaya are currently operating and the Atatürk dam is being built. Feasibility studies and the final designs of the Birecik and Karkamiş dams have been completed. Since the active storage capacity of these reservoirs will be 47.6 km^3 (1.6 times the annual mean flow of 30.5 km^3), the natural flow of this river will be regulated to a great extent by utilizing the head of 503 m from Lake Keban to the border over a distance of 468 km. Evaporation rates at the reservoirs in Turkey compared to those at Tabqa, Qadisiyah and Habbaniya are much less, due to the climatic conditions and improved volume-to-surface ratio of the Turkish reservoirs in the Euphrates valley.

On the other hand, the absence of large reservoirs in Syria and Iraq indicates that little practical use has been made of reservoirs in these countries for storing water from high-flow years to low-flow years, and flood waters will continue to flow to the sea.

The timing of the floods on the Euphrates and Tigris has never been ideal for crop production. As Garbrecht notes (quoted in Goldsmith and Hildyard 1984, p. 304):

First the floods of the Tigris and Euphrates were very erratic and occurred at the 'wrong time', the period April–June being too late for summer crops and too early for the winter crops. Secondly, the two rivers carried a much greater amount of sediment than the Nile River. And, finally, the very small incline of the alluvial plain [1:26,000] and the fine texture of soil easily gave way to waterlogging and salinization (lack of natural drainage).

The low-lying plains in Syria and Iraq form a natural expansion zone for high waters. The combined area of the lakes and swamps at the head of the Gulf varies from 8288 sq. km at the end of the dry season to 28,490 sq. km during the spring flood, covering the area having irrigation facilities. During the 1946 flood, the total inundated area reached 90,650 sq.km (Naff and Matson, 1984, p. 85), causing severe property damage and loss of life.

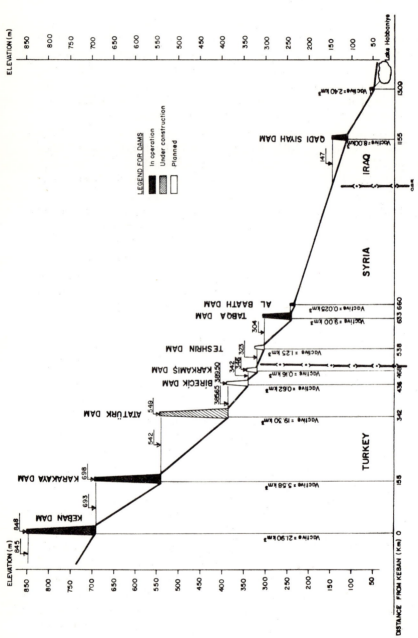

Figure 3. Profile of the Euphrates River (*Source*: DSL 1992).

Downstream riparian countries have no over-year water storage capacities. Therefore they are unable to store water for later use, as became clear during the long drought periods of 1958–62 and 1970–75. From the engineering point of view, potential reductions in natural flow at full development in the basin could be greatly mitigated by water savings provided by the regulation effects of the reservoirs in Turkey. On the one hand, Turkey's reservoirs could provide its neighbours with water security; on the other hand, the rate of siltation of downstream reservoirs would be much diminished.

Quantity parameters of a river can be transformed by storage reservoirs; in other words, the characteristics of a stream can be dramatically altered with the help of storage facilities. Such a change could be depicted in a flow-duration curve. For this purpose, a statistical analysis of the stream flow for the Euphrates at the Turkish-Syrian border was carried out with and without the regulation effect of the Keban dam. The annual run-off duration curves for the years 1937–90 for both cases are given in Figure 4 (DSI 1992). According to these curves, the frequency of mean annual flow rate of 968 m³/s, corresponding to 33 per cent of the time span, increased to 46 per cent after the construction of the Keban dam. Much higher rates are expected to be realized upon the completion of Atatürk dam.

Kolars (1993, pp. 13–14) asserts the positive implications of upstream regulations and points out that:

Variation in the flow of both rivers ranges from conditions of severe drought to destructive flooding, and it is on this basis that the *Turks make one of their strongest justifications for implementing the GAP with its giant dams and reservoirs* capable of smoothing out such variance and providing a dependable year-round flow downstream. However, this argument has not been persuasive enough for the Syrians and Iraqis. (Emphasis added.)

One of the most intensively impounded river systems in the world is the Colorado river which drains the south-western United States and enters into Mexico. A brief examination of the discussions which took place between the USA and Mexico provide us with an interesting insight into the run-off regulation within the context of the management of the entire basin.

At the time of negotiations on the Colorado river compact between the USA and Mexico, in view of certain allegations raised by Mexico, the USA's Department of State released the following statements on 30 June 1941 (Whiteman nd, pp. 947–8):

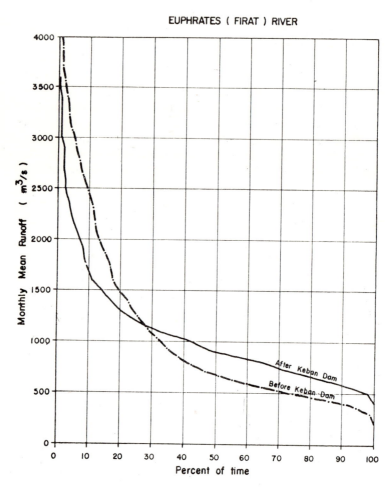

Figure 4. Run-off Duration Curve at the Border

The water it is proposed to deliver to Mexico from Colorado river in perpetuity is obviously worth many times a larger amount of uncontrolled normal and natural flow and hence would seem to be no less valuable than the 3,600,000 acre feet of normal and natural flow of water requested by Mexico in 1930. It is to be noted that there has been great variation in the annual flow of the river and that Boulder dam prevented serious shortages, even greater than those which would otherwise have occurred in 1937, 1939 and 1940. Moreover, the construction of the Boulder dam and the maintenance of expensive storage facilities and the water to be delivered to Mexico have not involved any cost to that country under the plan herein presented; no charge would be made to Mexico for storage costs at Boulder dam.

In the Department of State's memorandum of 11 February 1942, it was stated that:

. . . the Department of State felt that it had more than met the requirements of Mexico based upon that country's past claims since the quantity suggested of *controlled water would be so much more valuable than a much greater quantity of uncontrolled water*. It was noted with satisfaction that Mexico recognized this to a certain extent by its counter-proposal that approximately 2,000,000 acre-feet of water would be acceptable. . . . (Whiteman nd, pp. 948–9.)

These two memoranda clearly underline the importance of upstream regulation for basin-wide water resources management. It is interesting to note that, in the case of the Colorado, the annual volume of Colorado river water guaranteed to Mexico under the treaty of 1944, of 1,500,000 acre-feet (1,850,234,000 cubic metres), accounted for little more than 40 per cent of the 3,600,000 acre-feet of normal and natural flow requested by Mexico in 1930.

2.2. Water Transfer from the Tigris to the Euphrates

The total quantity of water flow in the Euphrates river regulated by large upstream reservoirs is likely to be adequate for domestic water supply, industrial growth and agricultural development in the foreseeable future; but there might still be a problem in matching the supply to the demand at certain places and times (e.g. during severe drought periods). To this end, Tigris river diversions seem to be technically, economically and hydrologically appropriate for the following reasons:

(i) Unlike the Euphrates, the Tigris river has several major tributaries in Iraq which enter the Tigris at the left bank from the Zagros mountains in the east. Among these tributaries are the Greater Zab, the Lesser Zab, the Adhaim and the Diyala. The average annual mainstream flow at Mosul is 23.2 km^3 and the tributaries supply a volume amounting to 29.5 km^3/year (Beaumont 1978). The total water resources of the Tigris basin, therefore, amount to 52.7 km^3/year, thus, 1.73 times as much as the annual mean flow of 30.5 km^3 in the Euphrates river.

(ii) According to the balance sheet of water resources versus water uses from the Tigris river prepared by Kolars (1992, p. 108), the amount of surplus water in the Tigris river is 11.9 km^3/year. In his balance sheet, Kolars accepts the natural flow as 49.2 km^3/year which is less than the figure of 52.7 km³/year given by Beaumont. Based on Beaumont's figure, surplus water amounts to 15.4 km³/year, of which 50 per cent could be transferred to the Euphrates. Topography in the Iranian part of the basin precludes the practical possibility of any significant water use there, or diversion to the other parts of Iran. Therefore, it is unlikely that Iran would be affected as a result of this transfer project.

(iii) In connection with this water transfer project, several authorities on Middle East water issues pointed out the important mutually supporting roles that both rivers play to each other. Some are quoted as below.

Iraq could well make greater use of the discharge in the Tigris. In fact, the Tharthar canal project which at the moment diverts Tigris Water into the Tharthar depression, thereby controlling floods, is planned to be extended to the Euphrates, facilitating therefore the transfer of flow from one river to the other (Anderson 1986, p.19).

The Iraqis are also planning to transfer water from the Tigris to the Euphrates. The Tharthar canal project presently diverts water into the Tharthar depression, controlling the flood flow of the Tigris. The next stage of the plan is a canal from the Tharthar into the Euphrates, and outlet canals back into the Tigris and Euphrates to channel water as needed into agricultural projects (Naff 1984, p.92).

Fortunately for Iraq, however, there is little suitable land in these two countries which could be irrigated by using the waters of the Tigris. As a result, it seems unlikely that serious international problems will be generated concerning the use of its waters, and Iraq will be able to make the fullest use of them for its own needs. This explains why Iraq is able to divert a significant proportion of the flow of the Tigris through the Tharthar Basin to augment the water resources of the Euphrates (Beaumont 1978).

Kolars (1993, p.49) makes a different recommendation concerning the route of a transfer canal, viz.:

. . . a canal might be built from the Mosul reservoir (or from a smaller retaining or diversion facility farther upstream) in order to bring a supplemental supply of water to the Euphrates river. Such a canal could run almost straight south following the 500-metre contour to the Euphrates below the Haditha Dam. This, in combination with water stored in reservoirs on the eastern tributaries of the Tigris, might alleviate Iraq's predicted water problems. The expenditure on such ventures should be considered as an international, regional item to be shared by all the riparians. Such an idea raises the possibilities of potential basin-wide/regional cooperation.

Another recommendation made by Beaumont (1991) is as follows:

On the Tigris the picture is clearer as much less development has occurred, or indeed little is planned outside Iraq. In Turkey some water use takes place in the Diyarbakir basin, but as yet no major water structure has been built, or seems likely to be built in the near future. Leaving Turkey, the river flows into Iraq, though for a short distance the boundary between Syria and Turkey is marked by the Tigris river itself. In this area the head waters of the Khabour, the major tributary of the Euphrates, are close by, and it would not be too difficult from an engineering point of view to divert some of the waters of the Tigris into the Khabour at this point.

Among the above-cited project proposals, the one which links the Tigris to the Euphrates through the Tharthar Valley has already been realized and operative since 1988 (Dhanoun 1988).

From time to time, it is argued that salinity in the Tharthar depression precludes the transfer of water except in extreme cases (Kolars 1993, p.13). However, a bypass canal to be built north of the Tharthar depression could transfer the fresh Tigris water directly into the Euphrates, by making use of the existing canal between the Tharthar depression and the Euphrates, avoiding the rather saline earth formation in the Tharthar lake bed (Figure 5).

While discussing the possibility of linkage between the Tigris and Euphrates rivers, it is interesting to note that the original idea dates back to pre-Christian times. It was then thought to link the two rivers by the Shatt el Hai canal (McDonald and Kay 1988, pp. 1–2).

This issue can be better put as follows:

Suppose two transboundary rivers enter into a lower riparian State. One of these rivers receives a large portion of its water from tributaries which run exclusively within national boundaries while the other river is highly susceptible to the demands of upper riparian countries. How ethical would it be for the lower riparian State to insist on maintaining all its existing and potential water rights on the latter river (which is very much needed and susceptible to depletions by other States) while reserving the surplus water of the former river only for itself?

A relatively similar case involving a water transfer was experienced by India and Pakistan. In 1954, the World Bank put forward a proposal for the equitable distribution of the water resources available to India and Pakistan. The proposal had three important features:

(i) The waters of western rivers were to be allocated to Pakistan and the waters of eastern rivers to India. Parts of Pakistan which were fed by

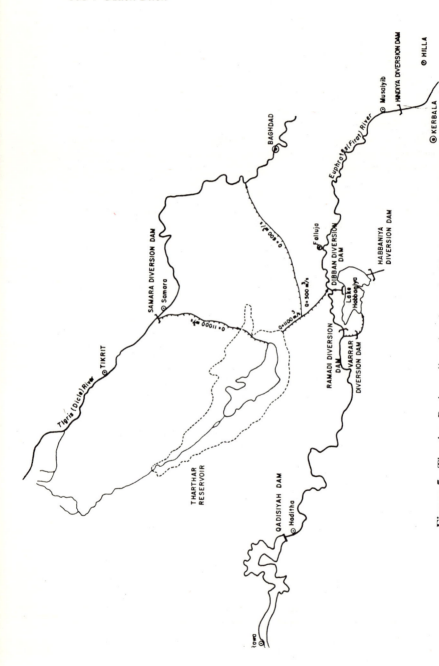

Figure 5. Tharthar Project: diversion from the Tigris to the Euphrates

the eastern rivers, would in future be fed by waters to be transferred from the western rivers by means of a system of link canals. It was estimated that 17.3 km^3 of water would be required, ultimately, to replace the water designed for use in India.

(ii) India would make a contribution to the cost of the replacement works.

(iii) During the construction phase, India would limit her withdrawals from the eastern rivers to proportions which would match Pakistan's capacity to replace (Framji et al. 1982).

The Bank's proposal differed from Pakistan's (which provided for existing uses to be supplied from existing sources), but it did recognize Pakistan's right to water in providing that India should pay the cost of building the replacement link canals. The gain to India would be that the waters of the eastern rivers would then be available for the expansion of irrigation in undeveloped Indian lands.

In sum, the Bank proposal protected existing irrigation uses from disturbance, and allocated surplus supplies to areas already developed or to be developed through water transfers among rivers. This was a technical solution which involved no judgment upon the legal contentions put forward by the concerned parties.

The India-Pakistan experience is of relevance in the Middle East. It illustrates that the existing and future agricultural water requirements in this region need not all continue to be met from the Euphrates. Some areas fed by the Euphrates could be more efficiently commanded by waters to be transferred from the Tigris river. A system of link canals can easily serve to augment the Euphrates-fed irrigation.

This possibility constitutes the most promising technical solution to help matching the supplies with the demands in the Euphrates-Tigris basin.

2.3 Environmental Issues

It is evident that water resources development projects have created some environmental problems. The goal of '*no damage to nature*', if strictly adhered to, would in some cases mean that developing or even using water resources might not be possible. However, economic development and environmental management can be concordantly pursued to minimize negative effects. This point was well made by McDonald and Kay (1988, p. 107):

There is a dichotomy between those who favour a technical fix to the problems of water supply and those who suggest that the technical solution will never solve this resource problem. . . . Such a rigid stance on this issue is unwise. To suggest that large-scale projects are not relevant to the needs of developing nations, where the water supply problems are huge, is clearly nonsense.

2.3.1 Impact on the Arabian-Persian Gulf

While dealing with the impacts of run-off regulation and irrigation schemes along the Euphrates and Tigris rivers, the dumping of the highly toxic trace metals and other forms of wastes into coastal and offshore waters by industrial plants should not be overlooked. Considering the fact that the Gulf countries have not yet reached a point of comprehensive marine management, we can anticipate even more dubious environmental consequences in the region in future. With or without irrigation development, under any circumstances, the Gulf area has been under serious attack by industrial and oil pollution, and agricultural pollution remains a trivial externality. The bottom line is that if one were to cite water resources development as the major cause of pollution in the Gulf, one would be missing the forest for a single tree.

2.3.2 Salinization and Return Flow Issues

It is a well recognized fact that the major part of arable lands in Iraq and Syria, including most of the present irrigated area, is seriously affected by salinization, and large areas have fallen out of production over the last few years. According to Tariq Harran (1973), Director General of Soils and Land Reclamation, in 90 per cent of the arable areas of central and southern Iraq, levels of salinity are so high that the average level of crop productivity per unit area in this region is below that in the majority of the Middle East countries. Indeed, Erik Eckholm describes vast areas of South Iraq which now 'glisten like fields of freshly fallen snow' (quoted by Goldsmith and Hildyard 1984, p. 140).

As for Syria, M. M. Gabaly (quoted by Goldsmith and Hildyard 1984, p. 140) noted that: 'Due to the aridity of climate, with evaporation exceeding precipitation in many locations it is estimated that 70 per cent of the soils put under irrigation are potentially saline.' Nonetheless, plans are afoot to irrigate a further 1.5 million acres as part of the giant Euphrates project. Annual crop losses due to salinity and waterlogging in the Euphrates valley alone already amount to $300 million. In short, we can conclude that all of the above-cited problems emerge from natural soil conditions and poor drainage.

On the other hand, the head-waters of Euphrates and Tigris are of high quality and the return flow from irrigation will be only moderately mineralized, containing about 700 ppm dissolved solids, and of satisfactory quality for irrigation supply (Lower Euphrates Project 1970). In this context we should note that under the terms of a joint treaty signed between Mexico and the United States, the USA agreed to reduce the salinity level of water entering Mexico to less than 800 ppm from an average salinity level of 2800 ppm at the Yuma desalinization plant (Goldsmith and Hildyard 1984, p.157). Thus, the agreed upon salinity level of return flow provided to Mexico is almost equal to that given by Turkey to its neighbours.

Moreover, the return flows from irrigation schemes around the Atatürk Dam enter directly into the dam reservoir and are diluted with large amounts of fresh Euphrates water. It is expected that the return flows may ultimately total 20 per cent or more of the diversions. This return flow is significant and is clean enough for additional irrigation in the downstream riparian countries.

Kolars (1993, p. 36) states that:

Syria may experience relatively little additional trouble regarding salination from Turkey, but its own soils are notoriously gypsiferous and saline and their proper washing and cleansing could dump oppressive loads of dissolved solids on Iraqi fields.

Although the lack of drainage facilities and the basic properties of soils are the major causes of salinization in arid climates, the salinization of soils is often solely attributed to quality of irrigation water. In this respect Kovda (quoted by Goldsmith and Hildyard 1984, p.147) makes the following point:

It has always underestimated the importance of the groundwater and properties of saline soils . . . secondary salinization of soils is attributed mostly to salts of irrigation water, which in fact are of secondary importance.

Conclusively, an efficient drainage scheme in the Euphrates-Tigris basin is of great significance, and the lack of drainage facilities is a major cause of several environmental problems, including salinization.

Issues of water quality as well as quantity in the Euphrates-Tigris basin, even under full development, are not more serious than those in any other developed countries' river basins (such as the Colorado), although doomsday scenarios are frequently drawn up for the future in this region.

2.4 Integrated Planning Concept of the Euphrates-Tigris Basin

With reference to the problems of transboundary rivers among riparian countries, the concept of integrated planning is merely presented in the context of resource allocation. However, the agreement on proper water allocation should be based on findings derived from a basin-wide planning process, and any negotiations should emphasize basin-wide planning as a goal. Such a plan depends on the collection, interpretation and evaluation of basic data relating to hydrology, climate, soils and other physical and socio-economic factors.

The presence of evident data anomalies in the available records concerning water and irrigable land resources in the Euphrates-Tigris basin have been noted several times in various reports, and the question of data validity is pertinent to the formulation of any firm conclusions. The current levels of extraction for irrigation and plans for development are not known with any precision.

Kolars (1993) points out that:

Understanding of the use of the Euphrates river and its tributaries in Syria for irrigation is obscured by lack of data and conflicting reports . . . much of the 640,000 ha originally scheduled for irrigation has had to be abandoned because of gypsiferous soils [p.42].

Early schemes to develop as many as 650,000 hectares along the Euphrates by building the ath-Thawrah Dam were reduced by 1983 to 345,000 ha and subsequently to 240,000 ha. Inaccurate soil surveys conducted by German firms failed to warn the Syrians about the effect of gypsiferous soils both on canals and on field applications of water. The Rasafah project originally estimated by the Russians to encompass 150,000 ha was actually abandoned and no more than 208,000 ha (12,000 ha government projects, 196,000 ha private lands) were under irrigation in the Euphrates valley in 1985–86. Moreover, large tracts of fertile valley land have been lost beneath the waters of Lake Assad and to poor drainage and salinization. Revisions in Syrian agricultural plans now place greater emphasis on dry farming and ancillary projects on the Khabur [p.9].

Naff and Matson (1984, p.97) noted that: 'Unexpectedly high reclamation costs of between $4000 and $10,000 per hectare had already led Syrian agricultural officials to admit privately that Tabqa's ultimate goal of 650,000 ha would probably never be reached.'

According to the USAID report quoted by Kolars (1991, p.8), less than half of the original 640,000 ha is reasonably good land for irrigation purposes.

According to Beaumont (1992, p.180), the actual amount of irrigation which is planned by Syria remains controversial, and figures have ranged from as low as 350,000 hectares to values in excess of 1 million hectares. Beaumont also adds, 'Recent estimates suggest that the final total will be between 400,000 and 800,000 hectares. . .' and he also points out, 'Iraq, too, has ambitious plans for irrigation expansion in the Tigris-Euphrates basin. Figures of in excess of two million hectares are quoted, but details are not available and it is not certain just how much of this proposed irrigation is to be located within the Euphrates catchment . . . .'

Based on the above quoted figures, Table 1 reveals data discrepancies on the existing and proposed irrigation project areas fed by the Euphrates river in Syria and Iraq.

Table 1. Conflicting Data on the Total Irrigation Project Areas Fed by the Euphrates in Syria and Iraq
(all figures in hectares)

Country Source	Syria	Iraq	Remarks
Official	773,000[1]	1,952,000[1]	
Kolars	375,000[2,*]	1,294,000[2]	*240,000 ha from
	397,000[3]	1,550,000[3]	the main stream plus 135,000 ha on the Khabur
USAID Report	320,000[4]		
Anderson	200,000[5] * to 500,000		*Figures include irrigation from the main stream
Beaumont	400,000[6] to 800,000		

Sources:

[1] Figures given to Joint Technical Committee (JTC) in 1982 and 1983
[2] Kolars (1991), pp. 8–10
[3] Kolars (1992), p. 107
[4] The USAID Report is not available to the author; quoted from reference 2 above, p. 8
[5] Anderson (1986), p. 18
[6] Beaumont (1992), p. 180

It is clear from Table 1 that a variety of local and foreign experts contend different figures concerning availability of irrigable land in each riparian country. Since irrigation is the major water consumer, a lack of consensus on irrigable land potential is an important issue. Such inconsistent figures can mislead analysts.

In conclusion, it can be seen that consistency and reliability of data on the land to be irrigated is a major concern for all parties and much work needs to be done to clarify the existing situation. Considering soil quality, soils are being classified in six categories ranging from excellent (class I) to poor (class IV) and to uncultivable (class VI). Among these categories, class IV has particularly severe limitations for crop production. High-textured soils, together with salinity and alkalinity, will cause serious difficulties in the process of reclamation, making it uneconomical. It is therefore not worthwhile to drain and reclaim such soils. Even after drainage and reclamation, the productivity of these soils would be very low compared to lighter-textured and better structured soils. Low productive soils, on which low yields are likely to be obtained despite enormous water use must be removed from irrigation in all riparian countries. Even if only a small percentage of the lands which are least suited for irrigation are removed from irrigation, the resultant water savings will be considerable.

Agricultural withdrawals from the Euphrates and Tigris, which correspond to 70–75 per cent of total consumption, are differently accounted for by the parties because of the soil data inconsistency mentioned above. National guidelines being practised by each country for data collection, evaluation and processing are based on different criteria and are not readily applicable to transboundary water courses. Data collection and survey of water and land resources need to be jointly performed by the riparian countries so as to acquire a basis for water allocation questions.

Based on the above considerations, a three-stage plan can be proposed in brief as follows:

(i) Inventory studies for water resources should include *inter alia*: unified measurement, data compilation, exchange of flow and meteorological data from agreed upon meteorological and gauging stations, correlation of flow data as appropriate, and extension of the short period records to generate longer period records in consistence with an acceptable level of data reliability. The following key gauging stations are proposed:

Key Stations	Turkey	Syria	Iraq
On the Euphrates	Belkisköy	Kadahya Abu-Kamal	Husabia (Hit) Nasiriya
On the Tigris	Cizre	—	Fishkhabow Mosul Kut

(ii) Inventory studies for land resources should include: unified classification of soils, and determining irrigation water requirements for projects in operation, under construction and planned by following the general rules of the '*Rapid Rural Survey Approach*' through visits to jointly selected project areas.

(iii) The two major stages very briefly described above, concerning water and land resources inventory studies, should be integrated into a comprehensive master plan, combining the riparian countries' resource management plans and including water transfer projects from the Tigris to the Euphrates. Based on this master plan, a simulation study can be carried out to develop water budget and allocation models among riparian countries.

In order to expedite cooperative efforts, the three-stage plan should be carried out in accordance with a time-table. During the period needed to put the plan into action and during the implementation of the projects based on this plan (such as water transfer from the Tigris, or modernization and rehabilitation of irrigation schemes) water withdrawals from the Euphrates might be limited. Since final allocations will be calculated on the basis of the plan itself, the utilization of waters will be adjusted according to the outcome of this plan.

The two extreme points of view, *absolute territorial sovereignty* and *absolute territorial integrity*, must be moderated by the concept of 'equitable and reasonable utilization' of a transboundary water-course and the obligation of a riparian country 'not to cause appreciable harm to other water-course states'.

Biswas (1993, p.21) draws our attention to the complexity of the relation between the principle of equitable utilization and the principle of obligation not to cause harm. However, there is always a way to challenge this complexity by means of well-considered technical approaches. In order to put the above-mentioned principles into practice, comprehensive basin-wide plans including water transfers should be used as a technical tool.

Until today, only the protection of downstream riparians' claims to priority use of waters has been sought by the international media, notwithstanding the consideration of a basin-wide integrated planning concept. Beaumont (1992) stressed this point in the following comment:

With irrigation water what seems to have happened up to the present is that international lawyers place too much emphasis on the rights of the 'downstream' states and not enough on those of the 'upstream' users. It is all too easy to ignore that in the case of the Euphrates almost 90 per cent of total flow of the river is generated within Turkey.

In this context, reference should also be made to a statement by McCaffrey, special rapporteur of the International Law Commission (ILC):

. . .a downstream State that was first to develop its water resources could not foreclose later development by an upstream State by demonstrating that the later development would cause it harm; under the doctrine of equitable utilization, the fact that a downstream State was 'first to develop' (and thus had made prior uses that would be adversely affected by new upstream uses) would be merely one of a number of factors to be taken into consideration in arriving at an equitable allocation of the uses and benefits of the water course. (McCaffrey, 1992.)

To this end, upstream riparians will have to utilize a reasonable portion of outgoing transboundary waters in future, although this might entail reducing the water consumption of downstream states.

3. CONCLUSIONS

The Euphrates-Tigris Basin provides a typical example of maldistribution of waters in time and space. However, this problem can be solved if some effort is made by all concerned. In this respect, two points for action have been identified:

(1) Run-off regulation, provided by upstream reservoirs, is of great importance.

(2) In matching supplies to the demands on the Euphrates, it is in the best interest of all riparians to reach a decision for the amount of water to be transferred from the Tigris to the Euphrates without too much delay.

As stated earlier, a reasonable and appropriate assessment of the amount of water each country needs from both rivers, depends upon the availability of complete and accurate information on the land and water resources of the Euphrates-Tigris basin, to be included in a basin-wide comprehensive master plan.

In spite of the doomsday scenarios envisaged by some experts, implementation of the solutions discussed here can contribute to peace and prosperity in the Euphrates-Tigris basin.

ACKNOWLEDGEMENTS

The author wishes to express his gratitude to the Staff of the General Directorate of State Hydraulic Works (DSI) of Turkey for their constructive inputs. His special thanks go to Mr Mehmet Çağil for his valuable support. The views expressed in this paper are strictly those of the author and do not necessarily reflect official views of any government organizations.

REFERENCES

Anderson, E. W. (1986) 'Water Geopolitics in the Middle East: the Key Countries'. Conference on U.S. Foreign Policy on Water Resources in the Middle East: Instrument for Peace and Development, CSIS, Washington D.C, 24 November 1986, pp. 18–19.

Bağiş, A. I. (1989) *G.A.P. Southeastern Anatolia Project—The Cradle of Civilization Regenerated.* A publication sponsored by Interbank, Istanbul, pp. 43–44.

Beaumont, P. (1978) 'The Euphrates River—An International Problem of Water Resources Development'. *Environmental Conservation*, Vol.5, No.1, The Foundation for Environmental Conservation, p.42.

—————— (1991) 'Transboundary Water Disputes in the Middle East'. International Conference on Transboundary Waters in the Middle East: Prospects for Regional Cooperation, Bilkent University, Ankara, 2–3 September 1991, p.15.

—————— (1992) 'Water—A Resource under Pressure', in Gerd Nonneman, ed., *The Middle East and Europe, An Integrated Communities Approach*, Federal Trust for Education and Research, London, p.180.

Biswas, A. K. (1993) 'Management of International Waters: Problems and Perspectives'. Middle East Water Forum, 7–10 February 1993, Cairo, p.21. (Revised version included in this volume).

Dhanoun, A. (1988) 'Tharthar Canal to be Opened'. *Baghdad Observer*, 4 January 1988, Baghdad.

DSI (General Directorate of State Hydraulic Works). 'Unpublished Statistical Compilation'. Investigation and Planning Department, Ankara, 1992.

Framji, K. K. et al. (1982) *Irrigation and Drainage in the World, A Global Review*, Volume 2. (ICID), p.1042.

Goldsmith, E. and N. Hildyard (1984) *The Social and Environmental Effects of Large Dams.* Sierra Club Books, San Fransisco, pp. 140, 147, 157, 304.

Harran, T. (1973) *Some Field Experimental Results of Salinity Problems in Iraq.* Ministry of Irrigation, D.G. of Soils and Land Reclamation, January 1973, p.1.

Kolars, J. (1993) 'Problems of International River Management: The Case of the Euphrates'. Middle East Water Forum, 7–10 February 1993, Cairo, pp. 9, 36, 42, 49. (Revised version included in this volume).

—————— (1992) 'Water Resources of the Middle East'. *Canadian Journal of Development Studies*, Special Issue, pp. 107–8.

—————— (1991) 'The Future of the Euphrates River'. The World Bank Conference, International Workshop on Comprehensive Water Resource Management Policy, Washington D.C., 24–28 June, pp. 8–10.

Lower Euphrates Project. *Feasibility Report*, Volume 3. Irrigation Development, Electro-Watt Engineering Services Ltd., Tipton and Calmbach Inc., Societe Generale Pour L'Industrie, Gizbili Consulting Engineers, April 1970, p. 241.

McCaffrey, S. C. (1992). 'The Law of International Water Courses: Some Recent Developments and Unanswered Questions'. *Denver Journal of International Law and Policy*, p. 509.

McDonald, A. and D. Kay (1988). *Water Resources Issues and Strategies*. John Wiley and Sons, Inc., New York, pp. 102–7.

Naff, T. and R. C. Matson (1984). *Water in the Middle East: Conflict or Cooperation?* A Westview Replica Edition, London, pp. 85–97.

Whiteman, M. M. (nd) *Digest of International Law*, Volume 3. Department of State, Files 711.12155/1915, 711.1216M/1199, pp. 947–8.

5 / The Jordan River and the Litani

MASAHIRO MURAKAMI and KATSUMI MUSIAKE

INTRODUCTION

The current basic framework for allocation of Jordan river water is enshrined in the 'Main Plan: 1953' and the 'Johnston Plan: 1955' which were negotiated with the United Nations but never formally endorsed by the governments concerned. The two plans basically agreed with the diversion from the Jordan river system to the coastal plain along the Mediterranean in Israel, but denied the integration of the Litani as a part of the Jordan system as proposed by Cotton in 1954.

Although these negotiations did not result in a formal agreement, both Israel and Jordan decided to proceed with water projects situated entirely within their own boundaries. Israel began work on the National Water Carrier in 1958 and completed the first stage work in 1964. It currently abstracts 90 per cent or more of the upper Jordan river flow from Lake Tiberias. The ambitious Israeli proposal for a Mediterranean–Dead Sea solar-hydropower project in 1980 was put aside, due to the strong opposition from the Arab states and others, and the drop in the world oil market price in 1984.

The Yarmouk river is the largest tributary of the Jordan system, with the largest unexploited potential except for a major part of the baseflow. The al-Wuheda dam scheme on the Yarmouk tributary, however, was halted in 1989 by strong opposition from Israel, which wanted more water in the Yarmouk river downstream. After Israeli occupation of Palestine in 1967, no multinational water projects on the Jordan river system could be promoted owing to political constraints with inter-state riparian questions.

It should also be recognized that issues of security of water resources and inter-state riparian questions in the Jordan river system have been some of the reasons why Israel could not withdraw from areas occupied since 1967, owing to Israel's heavy dependence on the recharge areas in the Arab catchment which include the extensive aquifers underlying the West Bank. Thus, without resolution of these inter-state water resources problems, no long-term peaceful settlement of the Palestine-Israel and Arab-Israel problems can be achieved.

THE JORDAN RIVER SYSTEM

The catchment of the Jordan river, excluding the upper basin, is all arid to semi-arid. Owing to this general aridity, a very large portion of the total area consists of endoreic or inland drainage. The total catchment area of the Jordan river is 18,300 km^2, of which 3 per cent lies in pre-1967 Israel as shown in Figure 1. The lower Jordan river between Lake Tiberias and the Dead Sea has a catchment area of 1050 km^2.

There is a marked spatial variation in the distribution of precipitation over the catchment. The recharge area is confined to the upstream mountainous areas of the Anti-Lebanon range where the mean annual precipitation amounts to 1400 mm, while the climate in the lower reaches of the Jordan river in the Rift Valley is arid to hyper-arid with an annual mean precipitation of less than 50–200 mm. The isohyets are shown in Figure 2.

The Jordan river originates in the south-western Anti-Lebanon range, on Mount Hermon, which is covered with permanent snow. The river flows through Lebanon, Syria, Israel and Jordan. The discharge that feeds the upper part of the Jordan river is derived principally from a group of karstic springs located on the western and southern slopes of Mount Hermon (Jabel Esh-Sheikh).

The river flows southwards for a total distance of 228 km along the bottom of a longitudinal graven known as the Rift Valley before emptying into the Dead Sea. Its principal tributary, the Yarmouk, forms the border between Syria and Jordan and divides Israel from Jordan in the Yarmouk triangle. The lower reaches of the Jordan river border on the Israeli-occupied West Bank to the west and Jordan to the east for a distance of about 80 km.

The quality of water in the headwaters of the north fork of the Jordan river is excellent with salinity less than 15 to 20 mg/l of chloride. The flow in the lower reaches of the system is supplemented by groundwater springs, but much of their contribution is so saline that it degrades the quality of the river flow, to the extent of several thousand parts per million of total dissolved solids (TDS) at the Allenby Bridge near Jericho.

Based on the nature of the hydrology, hydrogeology and water use, the Jordan river system may be classified into three sections, namely: (1) Upper Jordan river: Headwaters—Huleh Valley—Tiberias Lake; (2) Yarmouk river; (3) Lower Jordan river: Main stream—Dead Sea (Naff and Matson 1984).

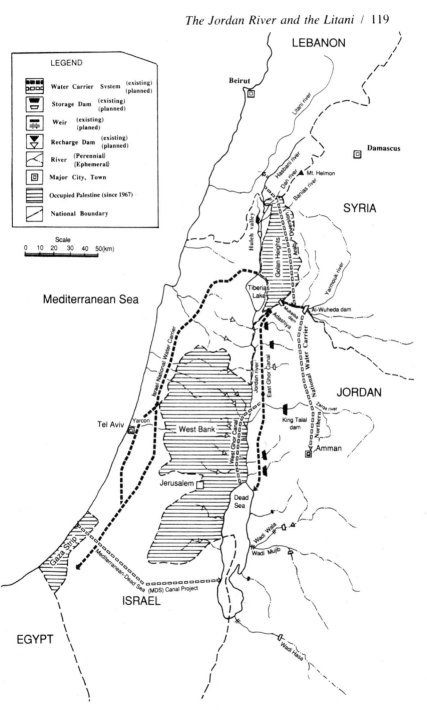

Figure 1. The Jordan River System

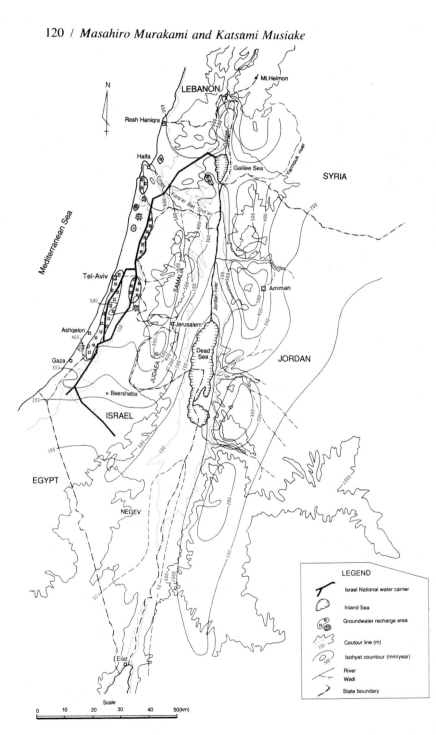

Figure 2. Mean Annual Rainfall and Israeli National Water Carrier System

1. The Upper Jordan River

The Upper Jordan river system includes three major headwater streams: (i) the Dan, (ii) the Hasbani and (iii) the Banias; and (iv) the Huleh Valley and (v) Lake Tiberias or the Sea of Galilee (Figure 1).

(i) *The Dan river* has the largest springs, which rise from Jurassic carbonate rocks, and supplies a large and relatively steady flow that responds only slowly to rainfall events. The average discharge of the Dan spring is 245×10^6 m^3 per annum, which makes up effectively the entire flow of the Dan river. The Dan spring is the least available in discharge among the major karstic sources of the Upper Jordan: its discharge varies from 173 to 285×10^6 m^3 per annum. The Dan typically represents 50 per cent of the discharge of the three rivers into the Upper Jordan.

(ii) *The Hasbani river* derives most of its discharge from two springs, the Wazzani and the Haqzbieh, the latter being a group of springs on the uppermost Hasbani. All of these springs rise from subsurface conduits in cavernous Cretaceous carbonate rocks. The combined discharge of these two springs averages 138×10^6 m^3 per annum, but the range of values measured varies over a greater range than that of the Dan spring. Over a recent twenty-year period, the flow of the Hasbani varied from 52×10^6 to 236×10^6 m^3 per annum. The Hasbani discharge responds much more rapidly to rainfall than does the discharge of the Dan spring.

(iii) *The Banias river* is fed primarily from the Hermon springs that issue from the contact of Quaternary sediments over Jurassic limestone in the extreme north-east of the Jordan Valley. The average discharge of the Hermon springs is 121×10^6 m^3 per annum: during a recent twenty-year period their discharge varied from 63×10^6 to 190×10^6 m^3 per annum.

In a typical year, these karstic springs provide 50 per cent of the discharge of the Upper Jordan river: the rest is derived from surface run-off directly after winter rainfall. In dry years, however, spring outflow may make up as much as 70 per cent of the flow of the Upper Jordan. The mean annual discharges of the three rivers are shown in Table 1.

Table 1. Mean Annual Discharge of Dan, Hasbani and Banias Rivers

Name of river	Mean annual flow (10^6m^3)	Annual range of flow (10^6m^3)	Riparian states
Dan River	245	173–285	Israel
Hasbani River	138	52–236	Lebanon
Banias River	121	63–190	Syria/Israel
Total	504	298–711	

The Dan spring, the largest of the sources of the upper Jordan, lies wholly within Israel close to the border with Syria. The spring sources of the Hasbani river lie entirely within Lebanon. The spring source of the Banias river is in Syria. These three small streams unite 6 km inside Israel at about 70 m above sea level to form the Upper Jordan river.

These spring systems together provide more water than can be accounted for as a result of rainfall over their immediate watersheds; thus, it is surmised that the springs represent the outflow of a large, regional aquifer. The combined outflow of the springs and the precipitation on the surface watershed of the Upper Jordan is of the order of 500×10^6 m³ per annum.

(iv) *Huleh valley* receives the flow from the Upper Jordan where it adds the flow of sublacustrine springs. Among the minor springs and seasonal watercourses contributing to the flow of the Upper Jordan, the most important is the Wadi Bareighhit. The water budget of the Huley Valley is shown in Table 2.

Table 2. Water Budget of the Huleh Valley

Source of Flow (10^6m³ per annum)	Inflow	Plus	Minus	Outflow
Flow into Huleh Valley	504			
Local runoff Huleh to Jisr Banat Yaqub		140		
Irrigation in Huleh Valley			-100	
Flow into Tiberias Lake				544

(v) *Tiberias Lake* lies in the centre of the Northern Great Rift Valley at 210 m below sea level. The Upper Jordan contributes an average of 660×10^6 m³ per annum to the lake, or 40 per cent of Israel's total identified renewable water resources. An additional 130×10^6 m³ per annum enters Lake Tiberias as winter run-off from various wadis and in the form of discharge from sublacustrine springs which have high salinity. The water balance of Lake Tiberias is shown in Table 3.

Lake Tiberias has a volume of 4×10^9 m³, which is 6.6 times the annual flow of the Upper Jordan inflow and 8 times the annual Jordan outflow. The water depth is 26 m on average, with a maximum of 43 m. The surface area is 170 km², which loses about 270×10^6 m³ per annum by direct evaporation.

The salinity of Lake Tiberias varies from a low value of 260 mg/l to a high of 400 mg/l of chloride, in which the variation depends primarily on the flow of the Upper Jordan with salinity less than 15–20 mg/l of

Table 3. Water Balance of Lake Tiberias

Source of Flow (10^6m^3 per annum)	Inflow	Plus	Minus	Outflow
Flow into Tiberias lake	544			
Rainfall over the lake		65		
Flow from local runoff		70		
Springs in and around lake		65		
Evaporation from lake surface			−270	
Outflow to lower Jordan				474

chloride. About 500×10^6 m^3 per annum leaves Lake Tiberias via its outlet, and flows south along the floor of the Dead Sea Rift for about 10 km to the confluence with the Yarmouk river.

2. The Yarmouk River

The Yarmouk river originates on the south-eastern slopes of Mount Hermon in a complex of wadis developed in Quaternary volcanic rocks. The main trunk of the Yarmouk forms the present boundary between Syria and Jordan for 40 km before it becomes the border between Jordan and Israel. Where it enters the Jordan river 10 km below Lake Tiberias (see Figure 1), the Yarmouk contributes about 400×10^6 m^3 per annum (Huang and Banerjee 1984).

There is no flow contribution from the part of the valley where Israel is a riparian. Of the 7242 km^2 of the Yarmouk basin, 1424 km^2 lie within Jordan and 5252 km^2 within Syria. The flow of the Yarmouk is derived from winter precipitation that averages 364 mm per annum over the basin (Naff and Matson 1984).

The Yarmouk River is the largest tributary of the Jordan river system, of which the potential resources have not been fully exploited except for a major part of the baseflow. The stream flow is supplemented by spring discharges from the highly permeable zones in the lavas; some further spring discharges may be channelled to the surface on wadi floors via solution pathways in the underlying limestone.

The mean annual flow discharge is 400×10^6 m^3 per annum, which is 65 per cent of the total discharge of 607×10^6 m^3 per annum of the East Bank of Jordan. The flow is largely influenced by rainfall pattern in the Mediterranean climate, indicating a maximum monthly discharge of 101×10^6 m^3 in February and a minimum of 19×10^6 m^3 in September. The water salinity of the Yarmouk river is quite low, being in the range between 280 and 480 mg/l of total dissolved solids (Huang and Banerjee 1984).

3. The Lower Jordan River and Dead Sea

South of its confluence with the Yarmouk, the Jordan flows over late Tertiary rocks that partially fill the Rift Valley. For the first 40 km the river forms the international boundary between Israel and Jordan; south of that reach, it abuts the Israeli-occupied West Bank of Jordan, where it forms the present cease-fire line. The Jordan here flows through the deepest portion of the Rift Valley to enter the Dead Sea at 401 m below sea level, the lowest point of the earth.

Run-off from winter rainfall within the valley is carried to the Jordan river via steep, intermittent tributary wadis incised in the wall of the Jordan valley, primarily on the East Bank. The source represents an additional 523×10^6 m^3 of water per annum, of which only 20 per cent originates in Israel; 286×10^6 m^3 per annum is derived from perennial spring flow, while 237×10^6 m^3 per annum is provided by winter rainfall (Naff and Matson 1984). There are nine major wadis with perennial flows in the eastern Jordan valley: wadi Arab, wadi Ziglab, wadi Jurm, wadi Yabis, wadi Kufrinja, wadi Rajib, wadi Zerqa, wadi Shueib, and wadi Kafrein. The mean annual flow discharge of the nine wadis is preliminarily estimated to be 590×10^6 m^3 per annum in total (Huang and Banerjee 1984), as shown in Table 4.

The quality of the lower Jordan river is influenced both by rainfall patterns and by the amount of baseflow extracted upstream. Water salinity is about 350 mg/l of total dissolved solids (TDS) in the rainy season, while it rises to 2000–4000 mg/l in the dry season at Allenby bridge near Jericho.

Finally, the salinity of the Jordan river system reaches 250,000–300,000 mg/l of TDS in the Dead Sea, a level approximately seven times as high as that in the Mediterranean Sea, which is 40,000 mg/l. The Dead Sea salinity level is too high to sustain life, but certain minerals such as potash and bromines can be extracted by solar evaporative processes.

The Dead Sea covers an area of 1000 km^2 at a surface elevation of 400 m below mean sea level. It has two basins separated by the Lisan Straits: the northern basin with an area of 720 km^2, and the southern basin with an area of 230 km^2. The 40,000 km^2 catchment area includes parts of Israel, Jordan and Syria. The shortest distance between the Dead Sea and the Mediterranean Sea is 72 km.

The Dead Sea is a closed sea with no outlet, except for very high evaporation from the sea surface which amounts to 1600 mm per annum. In the past, the evaporation losses were replenished by an inflow of fresh water from the Jordan river and its tributaries, as well as other sources such as wadi floods, springs and rainfall.

Table 4. Estimated Annual Flows in Eastern Jordan Valley and Dead Sea Basin
(Unit: $10^6 m^3$ per annum)

River system		Storm run-off	Base flow	Total
Eastern Jordan Valley				
Yarmouk river (Adashiya)		182.0	218.0	400.0
Wadi Arab		6.5	24.9	31.4
Wadi Ziglab		2.2	8.3	10.5
Wadi Jurum		0.2	11.5	11.7
Wadi Yabis		1.6	6.2	7.8
Wadi Kufrinja		1.0	5.8	6.8
Wadi Rajib		1.3	3.0	4.3
Wadi Zerqa		46.5	48.3	94.8
Wadi Shueib		1.8	8.0	9.8
Wadi Kafrein		1.4	12.0	13.4
	Sub-total	(244.5)	(346.0)	(590.5)
Dead Sea Basin				
Wadi Zerqa Ma'an		3.0	20.0	23.0
Wadi Wala		16.6	20.1	36.7
Wadi Mujib		20.7	21.0	41.7
Wadi Al-Kerak		3.2	15.0	18.2
Wadi Hasa		4.9	36.3	41.2
	Sub-total	(48.4)	(112.4)	(160.8)

There are five major wadis with perennial flow in the eastern Dead Sea basin, including wadi Zerqa Ma'an, wadi Wala, wadi Mujib, wadi Kerak, and wadi Hasa. The mean annual discharge flowing directly into the Dead Sea is preliminarily estimated to be 160×10^6 m³ per annum in total (Huang and Banerjee 1984) as shown in Table 4.

The mean volume of water flowing into the sea before 1930 was about 1.6×10^9 m³ per annum, of which 1.1×10^9 m³ per annum were carried by the Jordan river (Weiner and Ben-Zvi 1982). Under these conditions, the Dead Sea had reached an equilibrium level at around 393 m below sea level, with some seasonal and annual fluctuation due to variations in the amount of rainfall. However, since the early 1950s, Israel, and later on Jordan, have taken steps to utilize the fresh water flowing into the Dead Sea for intensified irrigation and other purposes, which has reduced the amount of water entering the Dead Sea by 1×10^9 m³ per annum. Consequently, the water level in the Dead Sea has declined in recent years, reaching as low as 403 m below sea level today, which is almost 10 m lower than its historic equilibrium level as indicated in Figure 3. The surface area of the Dead Sea and its evaporated volume vary only by a few percentage points between elevations from −402 to −390 m, while water levels fluctuate considerably.

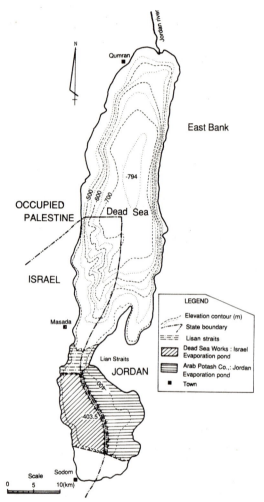

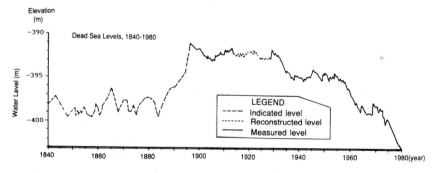

Figure 3. Dead Sea and Historical Change in Water Level

4. Riparian Questions of the Jordan River System

In 1953, the four countries, Lebanon, Syria, Israel and Jordan, agreed basically upon the priority use of Jordan river waters, in the so-called 'Johnston Agreement', including the priority use of the main stem of the Jordan river by Israel and Lebanon. The biggest tributary of the Yarmouk river which runs along the northern border of Jordan with Syria would be exclusively used by Syria and Jordan. This established a water allocation of the usable Jordan river estimated at 1380×10^6 m^3 per annum in total; 52 per cent (720×10^6 m^3) to Jordan, 32 per cent (440×10^6 m^3) to Israel, 13 per cent (180×10^6 m^3) to Syria and 3 per cent (40×10^6 m^3) to Lebanon (Naff and Matson 1984). It is widely assumed that the technical experts of each country involved in this discussion agreed upon the details of this plan, although soon afterwards the governments rejected it for political reasons.

With the failure of these negotiations, both Israel and Jordan decided to proceed with water projects situated entirely within their own boundaries. As a result Israel began work on the National Water Carrier in 1958 which is currently abstracting 90 per cent or more of the flow of the Upper Jordan river by the intake at Eshed Kinrot on the north-west shore of Lake Tiberias.

Syria has implemented a number of small–medium size dam development schemes for the Upper Yarmouk. These schemes have all contributed to increased salinity levels in the Lower Yarmouk and Lower Jordan rivers, lower water levels in the Dead Sea, and reduced irrigation water for Jordan's East Ghor Development Project. From a strategic point of view, this long-term Syrian effort could reduce Jordanian access to the Yarmouk, on which Jordan relies to irrigate the Jordan valley, and may affect downstream availabilities for Israel. This could conceivably lead to the possibility of heightened tension or even armed conflict among the riparians (Starr and Stoll 1987). In 1988, Jordan and Syria signed a protocol of understanding which paves the way for the commencement of al-Wuheda dam on the international border of the Yarmouk River, but this accord was never discussed and/or negotiated with the Israeli government.

Construction of the diversion tunnel of the Al-Wuheda dam could not be continued owing to opposition from Israel which still seeks an increase, rather than a decrease, of the Yarmouk's flow downstream.

THE LITANI RIVER AND RIPARIAN QUESTIONS

The World Zionist Organization had demanded in 1919 that the east-west section of the lower reaches of the Litani river become the international border of Palestine. The Lowdermilk Plan in 1944 treated the Litani as a part of the Jordan river system and proposed diverting 40 per cent of Litani flow into the Upper Jordan. This diversion plan aimed to generate hydroelectricity before supplying water to Lake Tiberias for further use through a water carrier system. An engineering scheme was designed by Cotton in his Master Plan of 1954 on the basis of regional cooperation between the two states. This interpretation was not accepted by the American negotiator Eric Johnston who resolved the discrepancies between the Cotton Plan (1954) and the Arab Plan (1954), and finalized a unilateral development plan in 1955 as indicated in Table 5.

Table 5. Water Allocations to Riparians of Jordan River System
(Unit: $10^6 m^3$ per annum)

Plan	(Source/Year)	Lebanon	Syria	Jordan	Israel	Total
Main Plan	(UN:1953)	—	45	774	394	1213
Arab Plan	(Arab:1954)	35	132	698	182	1047
Cotton Plan	(Israeli:1954)	451	30	575	1290	2346
Johnston Plan	(USA:1955)	35	132	720	400	1287

1. The Basin Hydrology and Water Resources Development

The Litani river lies entirely within the national borders of Lebanon. The flow of the Litani varies from year to year with an average of approximately $700 \times 10^6 m^3$ per annum, of which only $60 \times 10^6 m^3$ is contributed below Nabatiya. Of the annual flow, 60–65 per cent occurs during winter, from January through April; 15 per cent occurs during May and June; 12 per cent from July through October; and 10 per cent during November and December.

Significant water resources development on the Litani requires the use of dams and reservoirs to regulate cyclical fluctuations. The Litani River Authority which was created in 1954, completed the main features of a series of projects by 1966, including Qirawn reservoir with a storage capacity of $220 \times 10^6 m^3$, and Awali hydroelectric power system. With the new hydroelectric system activated, the water resources of the Litani system were dramatically reallocated geographically. Substantial diversion of waters of the Litani were made to the Awali river by a tunnel conduit, making the Awali the largest river ($645 \times 10^6 m^3$/year) in Lebanon as shown in Figure 4. The flow diagram and water budget of the Litani is shown in Figure 5. This diversion for the hydro-power scheme leaves only $125 \times 10^6 m^3$/year of water for the lower Litani, with

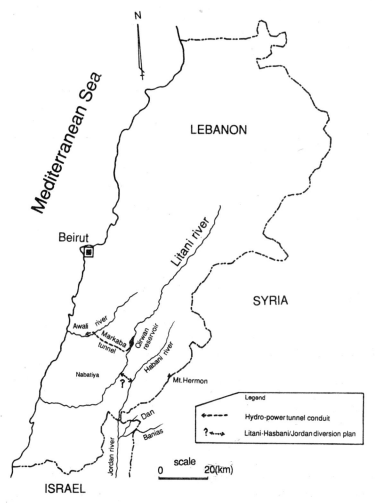

Figure 4. The Litani River System

essentially no water available in the five- to six-month summer period (Naff and Matson 1984).

2. Riparian Questions for the Inter-State Diversion Plan

The middle reaches of the Litani river near Nabatiya flow north to south parallel to the Hasbani river (upstream of the Jordan river) for less than 5–10 km, as shown in Figure 4. The Litani river water is a high quality water source averaging about 20 ppm of salinity. These features are

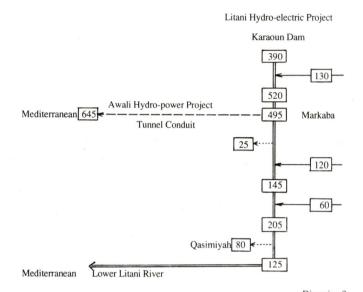

Figure 5. Water Budget of the Litani

the main attractions for Israel in wishing to divert the flow from the Litani to the Hasbani.

As noted above, large quantities of the Litani waters are diverted to the Awali. Accordingly, annexation of southern Lebanon and the seizure of the Litani waters have been frequently discussed at Israeli cabinet meetings, but have always been rejected for political reasons. The Lebanese understandably fear continued Israeli interest in the Litani, since Israel has attacked dams and other waterworks of its enemies in the past, including Nukheila dam and East Ghor Canal in the Yarmouk river and the Hasbani/Jordan river head-water diversion. When peace does come to Lebanon, and developmental pressures resume, there will be an increasing internal demand placed on the reserves of the Litani river.

BASIN WATER DEVELOPMENT ON THE EAST BANK OF THE JORDAN RIVER

The Hashemite Kingdom of Jordan carried out water resources development in the Yarmouk river and East Jordan valley, including three major schemes: (i) East Ghor Main Canal to divert the baseflow of the Yarmouk river, (ii) Side-wadi dams to store the stream flows in the East Jordan valley, and (iii) al-Wuheda dam to store the flood discharge of the Yarmouk river.

1. East Ghor Main Canal (EGMC) Project

The Yarmouk river, which is the largest tributary in the Jordan river system, has a mean discharge of 400×10^6 m^3 per annum. The Yarmouk provides almost half of the Jordan's surface water resources. The water in this river, after allowing for some 17×10^6 m^3 per annum for downstream users in Israel, is diverted through the East Ghor Main Canal (EGMC), an irrigation canal which runs along the Jordan river, to serve agricultural water needs in the Jordan valley. The upper East Ghor Canal phase was completed in 1964, and it had reached a length of 100 km by 1979, which permitted irrigation of 22,000 ha in total (Beaumont et al. 1988) after raising dam heights at the end of the 1980s.

Shortage and/or limitation of groundwater resources to meet the growing municipal and industrial water demands in North Jordan necessitated conveyance of 45×10^6 m^3 per annum of water from EGMC to Amman by pumping the extremely high head of 1300 m from the Deir-Alla treatment and pumping station (-200 m below sea level) to the terminal reservoir ($+1100$ m above sea level). The schematics of water transport systems in North Jordan are shown in Figure 6. Due to required priority for the irrigation sector, the system is not allowed to supply water during the summer season, and consequently only about 28×10^6 m^3 of water are being pumped per annum.

2. Side-wadi Dams

The King Talal dam on the Zarqa river was completed in 1979, to collect not only natural flows in the river system but also sewage effluents, both treated and untreated, from the population centres of Amman and Zarqa. An increasing proportion of the water stored in the dam comprises sewage effluents, of which the amount of treated sewage is expected to increase from 29×10^6 m^3 in 1985 to 116×10^6 m^3 in 2005 and 165×10^6 m^3 in 2015 in the Northern Jordan (World Bank 1988). Although the quality of water in the reservoir is still good, and suitable for cultivation of most crops (except for leafy vegetables) through drip irrigation, the use of King Talal water for municipal and industrial (M & I) purposes, even after treatment, is to be avoided, taking into account the health risks involved.

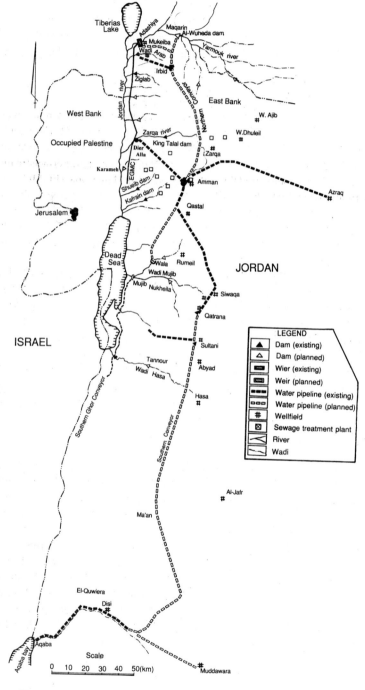

Figure 6. Water Resources Development Projects in Jordan

Small embankment-type dams of between 30 and 38 m have been constructed since 1968 in several rift-side wadis, including Ziqlab dam with 4.3×10^6 m³ of storage, Shueib dam with 2.3×10^6 m³ of storage, and Kafrain dam with 3.8×10^6 m³ of storage. The rift-side dam scheme on wadi Shueib was intended to store winter flows for downstream irrigation, but it has not been able to do so effectively, owing to substantial leakage through its gravel foundation and the limestone geology in and around the reservoir.

Wadi Arab dam, which was completed in 1987, has a total storage capacity of 20×10^6 m³ at a cost of U.S.\$50 million. The scheme was originally planned to store 30×10^6 m³ per annum of spring flow in the wadi; however, the flowing spring suddenly stopped owing to groundwater development in the adjacent well-field in the wadi in 1985. The dam design had to be amended to store an excessive winter flow from the EGMC by pumping up 100 m for M & I water supply during the summer season. This was made possible by raising the dam height and changing the supply objectives to M & I. The combined capacity of the King Talal and Wadi Arab dams had been increased up to 130×10^6 m³ by the end of the 1980s.

3. al-Wuheda Dam

The al-Wuheda dam, first conceived in 1956 to store the waters of the Yarmouk river, could have soon been constructed to the east of Maqarin about 20 km north of Irbid, as shown in Figure 6. The estimated streamflow at Maqarin gauging station is 273×10^6 m³ per annum on an average, which includes the flood waters being discharged downstream unused. Based on an agreement between Syria and Jordan in 1988, preliminary work on opening an 800-metre-long diversion tunnel was completed by the end of 1989. Its reservoir would have had a gross capacity of 225×10^6 m³ with an effective storage volume of 195×10^6 m³ annually. The water would have irrigated an additional 3500 hectares in the Jordan valley, and supplied 50×10^6 m³ a year to the Greater Amman area and Eastern Heights. It was also planned to generate an average of 18,800 kWh of electricity a year. Syria would have used part of the water and 75 per cent of the total hydroelectric power. The project was stopped, however, by strong opposition from Israel which sought more water from the Yarmouk river downstream.

BASIN WATER DEVELOPMENT IN ISRAEL

Israel commenced water resources development of the Upper Jordan river system soon after the establishment of the state in 1948. The initial stage of development was to improve the flow in the marshy Huleh valley, taking into account the lack of capital in the new state. The largest water resources development project in the Upper Jordan river system has been the 'Israeli National Water Carrier' which is a huge aqueduct and pipeline network carrying the waters of the Upper Jordan river southwards along the coastal plain.

1. Huleh Valley Drainage Work

The first stage of drainage of the Huleh swamps was begun in 1951 by Jewish rural settlers. The Huleh valley, which is situated in the northern-most corner of Israel, was a marshy area before the 1950s, where nobody could live. The marshy area was flooded by the winter flow of the Upper Jordan river, which then evaporated in the semi-tropical climate without productive use. The land reclamation works were performed by immigrants to provide a series of canals and drains to control both flood water and the groundwater levels in the depressions, to convert the valley from a useless marsh into fertile irrigation land. However, work on the Huleh swamp, which fringed on the demilitarized zone with Syria, provoked a number of military incidents.

Development of the upper river basin by irrigation and drainage of the Huleh valley, however, increased both saline and nutrient flows into Lake Tiberias and resulted in a heightened concern over eutrophication. The chloride ion concentration in the waters of Lake Tiberias (Sea cf Galilee) rose from below 300 mg/l to nearly 400 mg/l between the years 1950 and 1964.

2. National Water Carrier System

In the early 1950s, discussions took place between Israel and the neigh-bouring Arab states in an attempt to reach an understanding as to how the waters of the River Jordan might be mostly fairly allocated between the four states. This plan, which was drawn up for the United Nations, is usually referred to as the 'Main Plan: 1953'. After prolonged negotiations, modifications to the original plan were made and this new version became known as the 'Johnston Plan: 1955' after the American mediator, Eric Johnston. Total usable water in the River Jordan was estimated to be 1287×10^6 m^3 per annum, of which 31 per cent was to be allocated to Israel; however, the governments failed to reach any settlement for political reasons.

As a result, Israel began work on its National Water Carrier in 1958, and completed in it 1964. The main storage reservoir, and also the starting point of the scheme, is Lake Tiberias. From here water is pumped through pipes, from 210 m below sea level, to a height from which it flows by gravity into a reservoir at Tsalmon. After a further lift, the water flows via a canal to a large storage reservoir at Beit Netofa, which forms a key part of the system. South of Beit Netofa, the water is carried in a 270-cm-diameter pipeline to the starting point of the Yarqon-Negev distribution system at Rosh Ha'ayin. In the initial stages 180×10^6 m³ per annum of water were carried in 1964. This capacity was soon increased to 360×10^6 m³ per annum in 1968, and it is now believed that the maximum capacity approaches 500×10^6 m³ per annum (Beaumont et al. 1988). This has, however, not yet been attained owing to water salinity constraints in Lake Tiberias. At the present time, the national water grid interconnects all the major water demand and supply regions of the country, with the exception of a number of desert regions in the south. In total, it supplies approximately 1400×10^6 m³ per annum, or about 90 per cent of all Israel's water resources. More than half of the water is obtained from the Jordan river and its tributaries, with a further 14 per cent from the Yarqon river basin.

3. Israel's Occupation Policy and Water Resources of the West Bank

The occupied lands, most notably the West Bank and Golan Heights, are important to the water economy and security of Israel. One-third of Israel's water comes from the Jordan River. The 1967 conquests are important in this light also because the Golan Heights afford control over the Upper Jordan, enabling Israel to block any Arab attempt to divert its headwaters (see Figure 1). Almost half of Israel's total water supply therefore consists of water that has been diverted or pre-empted from Arab sources located outside its pre-1967 boundaries (Naff and Matson 1984).

MEDITERRANEAN–DEAD SEA (MDS) CANAL PROJECT

Israel announced performance of a feasibility study on a sea water hydroelectric power generation project in the early 1980s, but this had been preceded by master plan studies over many years before 1980. The Mediterranean–Dead Sea Canal hydro-power project, as it was called, was proposed to exploit the 400 m elevation difference between the Mediterranean Sea (zero metres) and the Dead Sea (−402 metres) by linking the two seas.

1. Alternative MDS Canal Routes

The much-discussed Mediterranean–Dead Sea (MDS) Canal has been based on four main alter.ative canal routes as shown in Figure 7 and Table 6.

Table 6. Alternative MDS Canal Routes

Alternative	Intake-Outlet	Length (km)	Remarks
(a) Northern Route	Israel-West Bank	154	Unilateral plan by Israel
(b) Southern Route (1)	Gaza-Israel	100	Bilateral plan by Israel
Southern Route (2)	Israel-Israel	120	Unilateral plan by Israel
(c) Central Route	Israel-West Bank	72	Bilateral plan by Israel
(d) Aqaba Route	Jordan-Jordan	175	Unilateral plan by Jordan

The shortest conduits of the 'Central Route' and the 'Southern Route (1)' have the advantage of minimizing the constraints of both cost and environment. The 'Central Route' conduit would be 72 km long, including a 15-km section of open conduit and a 57-km of tunnel 5 m in diameter. The first 30-km section would cross Israeli territory, and the second 42-km section would traverse the West Bank (occupied Palestine). The minimum distance route option was, however, put aside for fear of possible saline water (sea water) leakage from the tunnel which could contaminate fresh groundwater aquifers underlying the Judaean mountains.

After considering 27 alternative conduit routes to connect the two seas, the 'Gaza–Ein Bokek' route with an 80-km tunnel length was selected in 1982 to minimize the capital cost. The selected route, however, would cross the occupied Gaza Strip as shown in Figure 7. For political reasons, an alternative route was considered which would move the entrance of the canal northwards into Israeli territory. This would add 60 million U.S. dollars to the cost, and 20 km to the planned 100-km length (WPDC 1980). However, even if political problems in the Gaza Strip could have been avoided, they would certainly have been encountered in Jordan which shares the Dead Sea with Israel and also extracts minerals such as potassium from it.

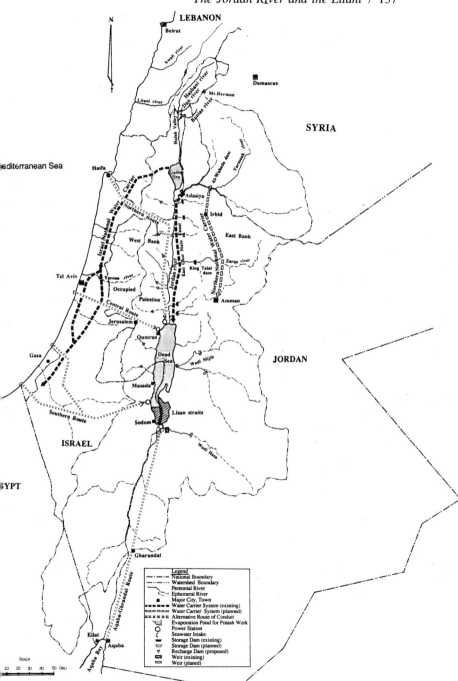

Figure 7. Mediterranean–Dead Sea (MDS) Canal Project

2. Israeli MDS Canal Plan

The Israeli solar-hydro development project, as the MDS canal, would generate 800 MW of electricity with annual generated electricity of $1.4-1.85 \times 10^9$ kWh, assuming a gross water head of 444–472 m and a maximum discharge of 200 m³/s with annual average flow intake of $1.23-1.67 \times 10^9$ m³ (Tahal 1982). The total project cost is estimated to be U.S.$ 1.89×10^9 (at 1990 prices) as shown in Table 7, assuming a 140 per cent price escalation from 1982 to 1990.

Table 7. Major Cost Elements of MDS Project

Cost Element	U.S.$ $\times 10^6$
Main tunnel ($= 80.4$ km)	732
Power station ($= 400$ MW $\times 2$)	385
Other facilities and structures	310
Design and supervision, etc.	142
Financial expenditure	319
TOTAL	1888

The planned effect of the canal was to raise the level of the Dead Sea by 17 m from 402 to 385 m below sea level. This would have meant that the mineral processing plants in both countries would have to be moved and potash production could fall by 15 per cent (WPDC 1980).

3. Jordan's Counter-Proposal for MDS

Jordan vied with Israel over the canal power scheme in 1981, by offering a counter-proposal to bring sea water from Aqaba Bay to the Dead Sea. This scheme would have also exploited the 400 m drop between ocean level and the Dead Sea to generate electricity. Sea water would have been pumped into a series of channels and reservoirs from Aqaba to Gharandal, 85 km further north (see Figure 7). From there, the water would fall into the Dead Sea to generate about 330 MW for 8 hours a day at peak demand (WPDC 1983).

4. Environmental Problems and Political Conflict

The flow of water from the Jordanian carrier would have forced Israel to cut back its own influx of water into the Dead Sea, or the level would have risen so high as to flood the potash works (of both Israel and Jordan) and the surrounding hotels on the Israeli side. The Mediterranean–Dead Sea hydro-power project was finally put aside, because of the strong opposition from the Arab states and others, and with the confusion and the drop in the world oil market prices in 1984. Israeli interest turned then to sea water pumped-storage from the Dead Sea (WPDC 1989).

It should be noted that a United Nations mission found that up to a level of -390.5 m the Dead Sea would not have flooded any religious or archaeological remains, would not have triggered earthquakes as this level was comparable with previous equilibrium levels, and would not have increased reflectivity. These studies therefore demonstrated that the project would not have had any adverse environmental effects (WPDC 1983). The possible increased evaporation through the introduction of Mediterranean water discussed below could have additional beneficial effects.

5. Dead Sea Pumped-Storage Scheme by Israel

Israel's Energy Ministry has recently shown renewed interest in a pumped-storage scheme on the Dead Sea, first proposed in the 1980s but shelved in favour of a similar project proposed for the Sea of Galilee. The power could be produced even more cheaply and efficiently with a pumped-storage scheme on the Sea of Galilee in northern Israel, but such a project could damage plant and animal life there. The interest has shifted back to the Dead Sea because of its almost total lack of flora and fauna. The Dead Sea pumped-storage scheme could produce 400–800 MW, equivalent to 8–16 per cent of the national grid's capacity of 5055 MW in 1990.

6. Co-generation Scheme for the MDS

The co-generation scheme was first conceived in the early 1980s to provide both hydroelectricity and fresh water with a reverse osmosis sea water desalination plant (Glueckstern 1982). The use of part of the hydro potential to make the reverse osmosis desalination cost-effective was put aside, however, owing to a poor understanding of the membrane technologies and the high cost at that time.

Discussion of the MDS scheme in the early 1980s may have overlooked the concept of shared resources and the benefits of joint development. Indeed up to now there has been no attempt to conceive comprehensive development of the Jordan river system including the linkage of the MDS and the al-Wuheda dam on the Yarmouk tributary. The proposed new co-generation approach to the MDS scheme thus takes into account: (i) recent innovative developments in membrane technology for reverse osmosis (RO) desalination which aim to save energy and to make desalination more cost-effective, and (ii) recent changes in the

Middle East political situation following the Gulf War that may make comprehensive basin development not only technically and financially feasible, but politically desirable and urgent.

(i) *Hydro-powered Seawater Reverse Osmosis Desalination for Co-generation*:

The co-generation scheme proposed would exploit the 400 m difference in elevation between the Mediterranean Sea and Dead Sea. The Dead Sea water would be maintained at a steady-state level with seasonal fluctuations of about 2 metres. This would sustain the sea water level between 402 m and 390 m below mean sea level, and the inflow from the Mediterranean should balance the evaporation.

The bilateral development plan for the Israel/Jordan Mediterranean–Dead Sea conduit scheme, hereinafter referred to as IJMDS, is a co-generation alternative which would combine the solar-hydro scheme with hydro-powered sea water RO desalination, as illustrated in Figure 8. The IJMDS scheme would have the following six main structural components:

(a) An upstream reservoir (the Mediterranean) at sea level, with essentially an unlimited amount of water.
(b) A water carrier and tunnel conduit with a booster pumping unit to lift water 100 m or more, or an open gravitational canal.
(c) An upper reservoir and surge shaft at the outlet of the water carrier to allow for regulating the water flow.
(d) A storage type hydroelectric unit capable of reverse operation to allow the system to also work as a pumped-storage unit, if required.
(e) A downstream reservoir, the Dead Sea, at a present surface elevation of approximately 402 m below sea level.
(f) A hydro-powered reverse osmosis (RO) desalination plant, including pre-treatment unit, pressure converter unit, RO unit, energy recovery unit, post-treatment unit, and regulating reservoirs for distribution.

(ii) *Estimates of Hydro-power Potential*:

The theoretical hydro potential of the head difference between the Mediterranean Sea (0 m) and Dead Sea (-400 m) by diverting 56.7 m^3/s (1.6×10^9 m^3 per annum) of sea water is estimated to be 194 MW. The hydro-power plant would generate 1.3×10^9 kWh per annum for the installed capacity at 495 MW with peak-power operation.

The Tahal plan of exploiting the gross head of 444–472 m (Tahal 1982) by transferring 43 m^3/s of sea water from the Mediterranean would have 198 MW of theoretical hydro potential. This alternative would generate

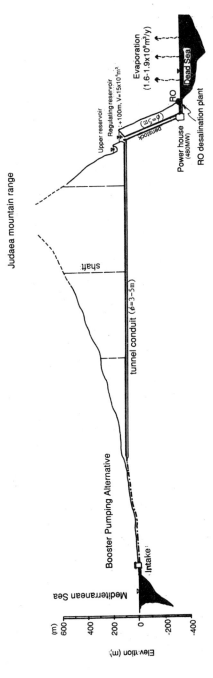

Figure 8. Joint Mediterranean–Dead Sea Conduit Scheme for Co-generation

generate 1.33×10^9 kWh for the installed capacity at 505 MW with peak-power operation.

These estimates of the hydro-potential are based on the conventional equations 1–4:

$$P_{th} = 9.8 \times W_s \times Q \times H_e \tag{1}$$
$$P = P_{th} \times E_f \tag{2}$$
$$P_p = P \times (24/8) \tag{3}$$
$$W_p = 365 \times 24 \times G_f \times P \tag{4}$$

where

P_{th} = theoretical hydro-potential (kW)
W_s = specific weight of seawater (−1.03)
Q = flow discharge (m³/s)
H_e = effective difference head of water (m)
P = installed capacity (kW)
E_f = synthesized efficiency (−0.85)
P_p = installed capacity for 8 hours a day of peak operation (kW)
W_p = potential power generation (output) per annum (kWh)
G_f = generating efficiency (−0.85)

(iii) *Methods of Co-generation for the MDS*:

The co-generation system is an application of solar-hydro-power to combine with hydro-powered sea water reverse osmosis (RO) desalination.

Booster pumping is necessary to make a head difference of 500 m, taking into account the estimated water pressure needed for the sea water reverse osmosis desalination. The sea water diversion capacity is estimated to be 50 m³/s, comprising 39 m³/s of intake water for hydro-power and 11 m³/s of feed water for the desalination unit.

The hydro-power unit would have a theoretical hydro potential of 160 MW, and would generate 1.2×10^9 kWh per annum for the installed capacity at 480 MW with peak-power operation for 8 hours a day.

The installed capacity of the RO plant to produce 100×10^6 m³ per annum of permeate is estimated to be 322,300 m³/day with load factor at 85 per cent.

The marginal operation of the RO system is designed to use the hydro-potential energy in a tunnel conduit (penstock) with 481.5 m of effective head of water for 16 hours a day during the off-peak time. The feed water requirements to produce 100×10^6 m³ per annum of permeate with 1000 mg/l of the total dissolved solids (TDS) are estimated to be 333×10^6 m³ per annum by assuming a 30 per cent recovery ratio. The amount of the brine reject is 70 per cent of the feed water. The brine reject of 233×10^6 m³ per annum has a salinity of 57,000 mg/l of TDS.

The brine reject is effectively used to recover the energy of the residual water pressure in the RO unit by a Pelton wheel turbine before discharging it into the Dead Sea. The energy recovery from the brine reject is estimated to be 24 MW with 135×10^6 kWh of annual electricity and load factor at 68 per cent. The recovered energy would be used to power the post-treatment process or generate electricity for other purposes.

(iv) *Cost Estimates of Hydro-powered Reverse Osmosis Desalination Plant*:

The project cost of the proposed hydro-powered sea water reverse osmosis desalination unit is preliminarily estimated to be U.S.$ 389,355,000 as capital and U.S.$ 44,387,000 per annum for O&M as shown in Table 8. The cost estimates are based on 1990 prices and the following assumptions:

- Plant life: 20 years
- Membrane life (replacement): 3 years
- Interest rate: 8 per cent during the three years of construction
- Cost benefit from energy recovery is not included.
- Costs for source water (groundwater) and pipeline/distribution are not included.

The unit water cost of the hydro-powered sea water reverse osmosis desalination for an annual 100×10^6 m³ of water is estimated to be U.S.$ 0.68/m³. This unit cost is about 20–40 per cent less than that of the estimated water tariff in the 'peace pipeline' project which is U.S.$ 0.85–1.07/m³ (Gould 1988).

The investment cost of the co-generation scheme is preliminarily estimated to be U.S.$ 2.3×10^9, including U.S.$$1.9 \times 10^9$ for the hydro-power unit and U.S.$ 0.4×10^9 for the reverse osmosis desalination plant.

(v) *Water Budget of the Dead Sea with MDS Conduit Scheme for Co-generation*:

Evaporation from the surface of the saline lake is the key factor in estimating the capacity for generating electricity by solar-hydro development of the Mediterranean–Dead Sea (MDS) Canal Project. For the same meteorological inputs and aerodynamic resistance, a decrease in salt concentration will increase evaporation rates and reduce lake temperature. From model analysis, which estimates the annual evaporation

Table 8. Major Cost Elements of RO Plant

Capital Cost Element	U.S.$(1990 price)
Pre-treatment	44,195,000
Desalting plant	70,414,000
RO membrane/equipment	84,835,000
Control and operating system	5,952,000
Appurtenant works	27,013,000
Powerline and substation	11,427,000
Energy recovery/turbine	2,999,000
Design and construction management	62,250,000
Financial expenditure	80,270,000
Sub-total	389,355,000

O&M Cost Element	U.S.$ per annum
Labour	3,718,000
Material supply	1,860,000
Chemicals	7,440,000
Power (booster pumping for RO feedwater)	3,100,000
Membrane replacement	28,269,000
Sub-total	44,387,000

rate and surface temperature as a function of aerodynamic resistance and thermodynamic activities of water (Calder and Neal 1984), the local evaporation rate is estimated to increase substantially by 345 mm per annum, amounting to an actual evaporation of 1908 mm per annum.

The Dead Sea surface, which is the source of evaporation for the MDS solar-hydro scheme, comprises two riparian states: Israel (300 km^2: 30 per cent) and Jordan (700 km^2: 70 per cent). The conduit route of the MDS scheme passes through the Gaza strips (10 km: 10 per cent) and Israel (90 km: 90 per cent).

The water budget for the Dead Sea co-generation scheme, to generate 1.2×10^9 kWh per annum of electricity and 100×10^6 m^3 per annum of fresh water, is as below:

- Evaporation after impounding seawater from the Mediterranean: -1900×10^6 m^3
- Tailrace water from MDS hydro-power station: $+1220 \times 10^6$ m^3
- Brine, reject water from RO plant: $+233 \times 10^6$ m^3
- Inflow from catchments: $+447 \times 10^6$ m^3

From a water budget study of the Dead Sea, the decrease in the inflow from the catchment will result in increasing the hydro-potential energy so as to introduce more sea water from the Mediterranean. The effort of decreasing the inflow to regulate the stream flow and control floods can be achieved by constructing a series of retention and/or storage dams in the catchment.

INTEGRATED BILATERAL BASIN DEVELOPMENT PLAN-21

The Dead Sea, which has huge hydro-solar (evaporation) potential, is shared by Israel and Jordan. The Dead Sea hydro-solar development for co-generation should therefore be discussed in the context of a master plan for inter-state development and management with the object of sharing of resources, and should provide the basis for peaceful collaboration between Israel and its neighbours.

Prior to elaborating on a new inter-state basin development plan for the Jordan river system, the following three major development alternatives are reviewed to identify the priorities for a master plan.

(i) *Inter-state water transportation by pipeline*: This comprises a 'Peace Pipeline', 'Euphrates–North Jordan Transmission' and 'Nile–Gaza/Negev Trans-pipeline' (see Figure 9). These three schemes were set aside, however, owing to fears about political constraints, including inter-state riparian right questions on the Euphrates and the Nile, where fears of water being used as a political weapon have been increasing. The Peace Pipeline project has now been emphatically rejected by all Arab states, who have stated that if necessary they will depend on non-conventional sources of water in their own territories, including sea water desalination. Both Israel and Jordan, which are not oil-producing countries, have been unable to adopt the thermal method of sea water desalination which requires substantial energy or electricity.

(ii) *Inter-state water transportation by tankers* could provide significant relief to all the coastal towns and cities of the Middle East. The provision of water by Turkey for the Israeli 'water-bag' scheme (see Figure 9) should go a long way towards developing Turkey's credibility and its good intentions as regards the Euphrates and Tigris, and will reduce one of the most serious problems for Israel in the discussion with the Palestinians. Turkey also holds the key to the future full use of the river systems of the Euphrates, Tigris, Ceyhan, Seyhan and Manavgat.

(iii) *Non-conventional water resources development* is likely to be given priority for the marginal waters in the non-oil-producing countries,

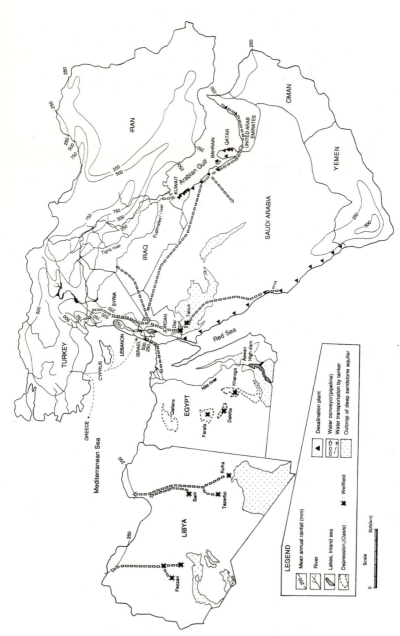

Figure 9. Water Resources Development Project Map of the Middle East

taking into account the difficulties not only in technical-financial-economical feasibility but also in political feasibility.

1. Integration of MDS Conduit for Co-generation and al-Wuheda Dam in Basin Development Plan-21

The Mediterranean–Dead Sea (MDS) conduit scheme for co-generation, which includes solar-hydro development as well as hydro-powered sea water reverse osmosis desalination, is a key proposal studied in this paper which aims at joint development by Israel and Jordan. Increased diversion of stream discharges in both the Jordan river and Dead Sea catchments, which are being lost to the Dead Sea, will result in increasing the sea water diversion capacity from the Mediterranean through a tunnel conduit. Three alternatives are shown below to reduce the discharges which are being wasted as they flow into the Dead Sea:

(a) al-Wuheda dam scheme on the Yarmouk river, which has been postponed since 1989, owing to the Israeli opposition on the question of downstream water allocation.

(b) Storage dam schemes in the riftside-wadis on the East Bank, including Wala and Nukheila dams on wadi Mujib and Tannour dam on wadi Hasa, which have no political constraints but need financial support from international agencies.

(c) Flood retention and groundwater recharge dam schemes in the side-wadis on the West Bank where limestone geology is predominant, are used to cut off winter flash floods which are being diverted into the Jordan river or the Mediterranean Sea, and/or to recharge the underlying aquifer system to sustain regional groundwater development. This may lead to an improvement in the difficult situation that Israel is facing, in which 40–50 per cent, or more, of its present water supply comes from an aquifer underlying the West Bank.

The al-Wuheda dam scheme with an effective storage capacity of 195 $\times$ 10^6 m^3, is Jordan's last major river development which is urgently needed to add 155 $\times$ 10^6 m^3 per annum of renewable fresh water to the national water supply grid. This will also reduce the amount of winter flow which is being lost to the Dead Sea. Meanwhile, to the west, the Jordan valley downstream of the al-Wuheda dam, including the Palestine region and a portion of Israel, needs more fresh water to extend irrigation development. To the south-east almost all the population centres in Jordan are located in highland deserts/at an elevation of 800–1000 m, which suggests the prior use of the Yarmouk river water for M & I water

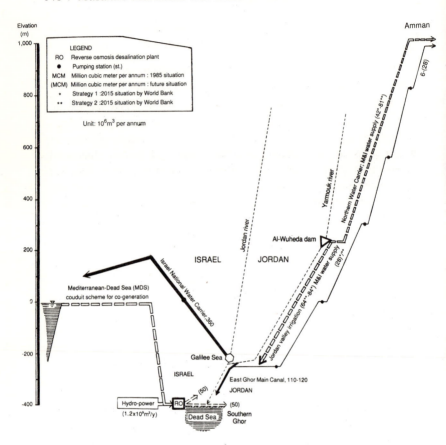

Figure 10. Schematic Profile of Integrated Joint Development Plan-21: al-Wuheda Dam and MDS Conduit Scheme for Co-generation

supply by diverting it from al-Wuheda (at 300 m) to Amman (at 800 m), as illustrated in Figure 10 showing the schematic profile of the Jordan Valley and Yarmouk river system.

An engineering proposal for the co-generation scheme, which combines solar-hydro development with hydro-powered sea water reverse osmosis (RO) desalination in the Mediterranean–Dead Sea (MDS) conduit scheme, generates the following two products, electricity and water:

– Electricity: 1.2×10^9 kWh per annum.
– Fresh potable water: 100×10^6 m³ per annum.

The MDS scheme for co-generation is also shown in Figure 10. The capital/investment costs of the hydro-power and the RO desalination are preliminarily estimated to be U.S.$ 1900 and 400 million, respectively. Annual potential outputs such as 1.2×10^9 kWh of electricity and 100×10^6 m^3 of fresh water for M & I supply from the co-generation system are each estimated to be equivalent with U.S.$ 80×10^6 per annum (U.S.$ 160×10^6 in total), assuming the tariff of electricity at U.S.$ 0.08/kWh and water at U.S.$ 0.8/m^3. These cost indices indicate that the MDS scheme is feasible for joint development by the two states.

The national power generation of Israel was 18.76×10^9 kWh per annum in 1988, about ten times as much as that of Jordan. Jordan's installed capacity of 480 MW was equivalent to 9 per cent of Israel's grid capacity of 5055 MW in 1990. The electricity from the Dead Sea hydro-power would be a resource to be shared by Israel and Jordan to supply their peak demands.

If current patterns of consumption are not quickly and radically altered, Israel, Jordan and the West Bank or Palestine will over-commit or deplete virtually all of their renewable sources of fresh water by the end of this century. In the circumstances, the Jordan river system, which comprises the al-Wuheda dam scheme on the Yarmouk river, unquestionably holds the greatest potential for either conflict or compromise. In the Southern Ghor of the Dead Sea catchment (the driest area of the Jordan Valley with annual rainfall less than 50–100 mm) there has been substantial demand for water to develop the region, but no alternative source of fresh water can be found in the area. M & I water demand in and around the Dead Sea by the year 2000 will be about 100×10^6 m^3 per annum, including increasing demands for mining (potash works), industry, agro-industry and resort hotels. The production of 100×10^6 m^3 of water per annum by hydro-powered desalination could be mainly used for M & I water supply with the aim of providing water exclusively for the low land in the Jordan Valley.

2. Method of Sharing and Allotment

Taking into account the water balance between the catchment flow system (inflow from rivers and wadis) and the Dead Sea (outflow by evaporation), all the dam schemes on the Jordan river system, including the Al-Wuheda dam and side-wadi dams, should be linked with the Mediterranean–Dead Sea (MDS) conduit scheme for co-generation in the context of an inter-state basin development master plan, to promote economic development for Israel and Jordan. The two riparians must

B. Uni-lateral Jordan river development (current situation)

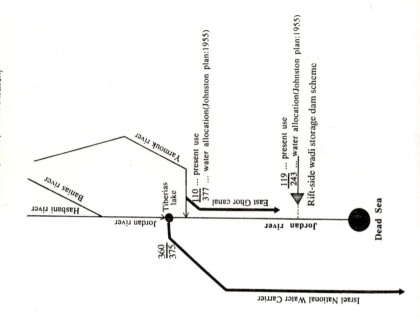

A. Water Allocation of Johnston Plan : 1955

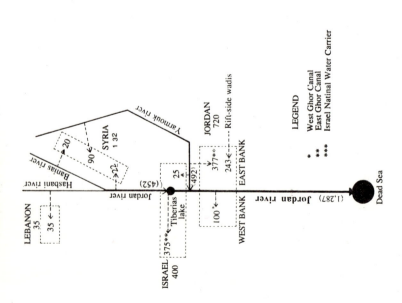

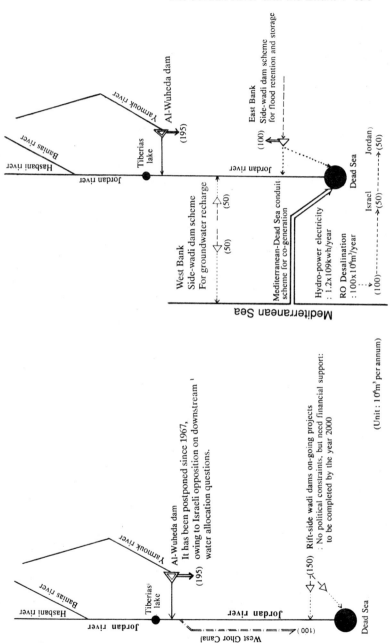

Figure 11. Flow Diagram of Water Allocation of Jordan River System

share the resources and benefits, and this may need further discussion of historical riparian questions and water politics.

Flow diagrams of water allocation of the Jordan river system are provided in Figure 11, including: (A) the Johnston Plan: 1955, (B) unilateral Jordan River Development which is the current situation, (C) ongoing or postponed projects which comprise political and/or financial constraints, and (D) proposed new schemes for integrated joint development plan-21, which include the MDS conduit scheme for co-generation and Side-wadi dam schemes for flood retention and groundwater recharge in the catchment.

The basic framework for a new inter-state Jordan river development plan-21 (for the twenty-first century) as conceived in this study would build on the 'Unified (Johnston) Plan: 1955'. The MDS conduit scheme with the new hydro-powered sea water reverse osmosis desalination would not only provide additional fresh water and clean energy (electricity) in the driest areas, but would promote integrated economic development between Israel and Jordan as a basis for lasting peace. A proposed bilateral basin development plan with water and energy allocation model-21 is shown in Table 9 and Figure 11 (D). This may influence the priority use and/or allocation of water and energy, including the following:

- al-Wuheda dam: use of 195×10^6 m^3 of water per annum in Jordan
- Side-wadi dams on the East Bank: use of 100×10^6 m^3 of water per annum in Jordan

Table 9. A Proposed Water and Energy Allocation Model-21
(Unit: 10^6m^3)

Source of Water	Lebanon	Syria	Jordan	Israel	Total
Unified (Johnston) Plan: 1955					
Hasbani	35				35
Banias		20			20
Jordan (main stream)		22	100	375	497
Yarmouk		90	377	25	492
East Bank Side wadis			243		243
Integrated Joint Basin Development Plan-21					
Water					
MDS hydo-powered RO desalination			50	50	100
Side-wadi dams in the West Bank			50	50	100
Electricity					
MDS hydo-power (10^9 per annum)			(0.14)	(1.06)	(1.20)

- Side-wadi dams on the West Bank: fifty-fifty use of 100×10^6 m³ of groundwater per annum in Palestine and Israel
- MDS co-generation for hydro-power: priority use of 1.2×109 kWh of electricity per annum, i.e. 1.06×10^9 kWh (88%) in Israel and 0.14×10^9 kWh in Jordan.
- MDS co-generation for desalination: fifty-fifty use of 100×10^6 m³ of fresh potable water per annum in Israel and Jordan/Palestine.

The total cost of implementing the proposed new joint basin development plan-21 is preliminarily estimated to be U.S.$ 4.5×10^9, including U.S.$ 2.0×10^9 for the MDS conduit hydro-power unit, U.S.$ 0.5×10^9 for the MDS sea water reverse osmosis desalination unit, U.S.$ 1.0×10^9 for the al-Wuheda dam and Side-wadi dams in both the East Bank and the West Bank, and U.S.$ 1.0×10^9 for operation and maintenance, and administration.

If cost sharing were to be on an equal basis between the two countries to assure the fifty-fifty benefit allotment, then project formulation including planning, design, construction, operation and maintenance, administration, and financing could be managed by an international consortium organized by an international agency such as the United Nations.

3. Remarks on the New Technology

This study was made to test the technical feasibility of exploiting sea water resources (including hydro-powered reverse osmosis sea water desalination for co-generation) by taking into account the distinctive nature of the arid zone hydrology and topography in and around the Dead Sea. Reverse osmosis is the cheapest process for desalination today, but it may not be the optimum solution for the twenty-first century. Further research will be needed to evaluate its technical feasibility, including: (i) the rate of actual evaporation from the Dead Sea surface after impounding, (ii) design of materials to avoid corrosion of hydraulic structures by sea water and brine or reject water, (iii) TBM methods of construction for the pipeline tunnel, (iv) application of low pressure (30–50 kg/cm²) type RO membrane modules for sea water desalination, (v) energy recovery efficiency in RO, (iv) methods of hybrid desalination, and (vii) power generation by solar ponds.

ACKNOWLEDGMENTS

The authors wish to express thanks to Professor Dr Yuzo Akatsuka of the University of Tokyo for his guidance and valuable advice. Thanks are also due to Professors Dr Hideo Nakamura, Dr Nobuyuki Tamai, Dr Sinichiro Ogaki and Dr Masahiko Isobe of the University of Tokyo for their comments. Masahiro Murokami is indebted to Nippon Koei Co., Ltd. for providing a research scholarship to the Institute of Industrial Science of the University of Tokyo since April 1990 which made this study possible.

REFERENCES

Beaumont, Peter, Gerald H. Blake and J. Malcolm Wagstaff (1988). *The Middle East: A Geographical Study* (2nd Edition), David Fulton publishers, pp. 100–105.

Calder, I. R. and C. Neal (1984). 'Evaporation from Saline Lakes: A Combination Equation Approach', *Journal of Hydrological Sciences*, Vol. 29, No.1, pp. 89–97.

Glueckstern, P. (1982). 'Preliminary Consideration of Combining a Large Reverse Osmosis Plant with the Mediterranean–Dead Sea Project', *Desalination*, vol. 40, pp. 143–56.

Gould, David (1988). 'The Setting of the Peace Pipeline', *MEED*, 26 March 1988, p. 10.

Huang, John and Arun Banerjee (1984). 'Hashemite Kingdom of Jordan: Water Sector Study, Sector Report', *World Bank Report* No. 4699-JO, 1984, pp. 35–6.

Murakami, Masahiro (1991). 'Arid Zone Water Resources Planning Study with Applications of Non-conventional Alternatives', Doctoral Thesis, Faculty of Engineering, University of Tokyo.

Murakami, Masairo, and Katsumi Musiake (1991). 'Hydro-powered Reverse Osmosis Desalination for Co-generation', International Seminar on Effective Water Use, Mexico 1991, International Water Resources Association (IWRA), pp. 688–95.

Naff, Thomas and Ruth C. Matson (1984). *Water in the Middle East: Conflict or Cooperation*, Westview Press, Boulder and London, pp. 17–27.

Starr, R. Joyce and C. Daniel Stoll (1987). *U.S. Foreign Policy on Water Resources in the Middle East*, The Center for Strategic and International Studies, Washington D.C., pp. 5–8.

Tahal (1982), 'MDS Project, Project Summary', Feasibility Study Report, Tel Aviv.

Weiner, Dan and Arie Ben-Zvi (1982). 'A Stochastic Dynamic Programming Model for the Operation of the Mediterranean–Dead Sea Project', *Water Resources Research*, vol.18, no.4, pp. 729–34.

World, Bank (1988). 'Jordan Water Resources Sector Study', *World Bank Report* No. 7099-JO, pp. 1–38.

WPDC, International News (1980), 'Israel Decides on Canal Route', *Water Power & Dam Construction*, October 1980, p. 4.

WPDC, International News (1983). 'Jordan attacks Dead Sea Project', *Water Power & Dam Construction*, March 1983, p.4.

WPDC, World News (1989). 'Dead Sea P-S Scheme Revived', *Water Power & Dam Construction*, May 1989, p.3.

6 / The Nile Basin: Lessons from the Past

YAHIA ABDEL MAGEED

1. INTRODUCTION

Emile Lodwig, the famous German historian-geographer, made the following remarks on the Nile when he visited Egypt and the Sudan in 1937: 'Every time I have written the history of man there hovered before my mind's eye the image of a river, but only once have I beheld in a river the image of man and his fate.'

Lodwig made this remark at a time of global confrontation, on the eve of the Second World War, which brought the threat of war to the Nile basin following Italy's occupation of Ethiopia. At the time, the whole basin was under the domination and influence of the European powers, and had been a stage for their play of rivalries since the beginning of the century.

Today in the post-cold-war era, times are equally uncertain. The confrontation of the big powers has been replaced by localized conflicts among the poorer nations of the world, and water is seen as one of the issues in these confrontations.

The Nile basin and its people are at the crossroads, facing a future full of risks and complexities of unprecedented dimensions. The population of the countries of the basin is expected to rise from 246 million to 800 million by the middle of the twenty-first century. There are scientific speculations that the basin is among the areas most threatened by global warming and sea-level rise, where one-fifth of Egypt's most populated and productive lands may be subjected to flooding.

The Nile basin occupies a prime location in the African continent (Figure 1). It overlooks the Mediterranean Sea in the north, and has the Red Sea and the Indian Ocean to its east. The renowned and fascinating River Nile witnessed spectacular ancient civilizations in its lower reaches in Egypt and no less important ones (Meroe and Axum) in its middle and upper reaches. To the ancient Egyptians the Nile became a holy river, God Hapi, bringing life and sometimes destruction to the most extensive arid desert land of the world. It attracted conquerers and invaders from Persia, Greece, Rome, Arabia and Europe. Over the years the basin became the stage for world politics and its riparian states became the victims of such politics.

Following the Turkish domination in the eighteenth century, the Nile

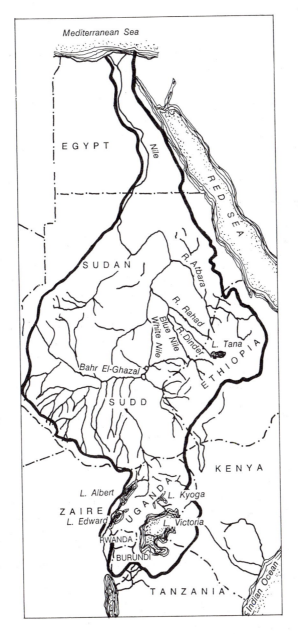

Figure 1. The Nile Basin

basin, because of its proximity to Europe and strategic location, became one of the main gates in the scramble for Africa. In 1884, the basin was divided into influence zones among the 'great powers' who met in Berlin: Britain, France, Belgium and Italy. This was followed by boundary agreements defining those zones: the 1891 Rome Protocol between Britain, Italy and France; the 1902 agreement between Britain and Menlik II of Ethiopia; and the agreement between Britain and the Free State of the Congo (Belgium). Most of the basin was under the influence of Great Britain, including Egypt, the Sudan and the East African countries: Kenya, Uganda and Tanzania. It is worth mentioning here that those boundary agreements included provisions to safeguard downstream interests in the flow of the Nile.

The colonial era and foreign interventions continued to the middle of the present century, and left their mark on present times. The colonial rulers greatly influenced the patterns of development in the basin for a long time. They deliberately worked to deepen the cultural, religious and ethnic differences, manipulating internal and inter-basin conflicts to divide and rule, and the Nile waters were often used as a tool. It is during the colonial era that basin-wide plans for the development of the Nile resources emerged, together with legal framework. The 1929 Nile water agreement between Egypt (which became independent in 1922) and Britain is one example.

After World War II, when all the basin states became independent, the Nile waters became an important issue in inter-state relationships. The basin became the scene of great power rivalries in the cold war era. The strategic position of the riparian states of the basin near the Mediterranean, the Red Sea and Suez Canal, and the Indian Ocean, made them relevant to the conflicts of southern Africa and the Horn of Africa, and between the Arab countries and Israel. Each of the nine countries of the basin has been at different times a target of Israeli economic political expansionism: it is worth mentioning here that early in the century there were attempts by Zionist circles to add part of the basin in Southern Sudan and Uganda to the list of possible Jewish settlements (Bashir 1986). The Nile water question has been a central issue in the power game and deepening rivalries between the riparian states and within states, as in the case of the High Aswan Project and the Jonglie Canal Project. During the post-war era, despite many initiatives for cooperation, the development of meaningful basin cooperation became futile, particularly between upstream and downstream states.

In the last two decades many parts of the basin have been affected by the persistent drought which struck the African continent from the Sahel in the west across the savannah belt in the Sudan to Ethiopia in the east. Even some pockets in the interior of the basin in the equatorial region could not escape the terrible effects of drought.

The climate crisis, coupled with the economic and debt crisis and the soaring population growth, brought to the basin a host of problems and grave consequences. Pockets of severe famine, mass immigration of rural poor to already strained urban centres, and migration across national borders due to internal and inter-state political unrest, led to grave environmental degradation of the basin's natural resources.

Many parts of the basin are also witnessing soil degradation and loss associated with erosion resulting from the over-exploitation of forests and destruction of vegetation cover. The situation is particularly severe in the Ethiopian highlands. The arid and semi-arid regions of the basin are now experiencing a serious breakdown of the environmental fabric and the spread of desertification along with a collapse of socio-economic systems.

The rainfed areas of the Nile basin, which provide the main sources of food for the basin population, suffer from problems of sustainability of production connected more with other factors than with water scarcity. These include the absence of land and water use policies, lack of infrastructure and other agricultural inputs, and deficient land and water management.

The huge irrigation systems that have spread all along the basin since the early twentieth century, as well as the medium and small systems that have been introduced in the upper reaches in reaction to the spells of drought, suffer from many constraints and problems.

In the lower reaches of the basin, water-logging and salinity problems are widespread as a result of poor irrigation and drainage combined with high cropping intensities. The upper reaches suffer from the absence of clear and consistent irrigation policies defining the role of irrigation in these areas, lack of experience, deficient capabilities and capital and financial constraints. All these have led to a decline in productivity and threats to sustainability.

Finally, deterioration in water quality is emerging as a threat to the basin water resources, with erosion in the upper catchments of the basin (particularly in the Ethiopian highlands) creating serious sedimentation in the lower reaches. On the other hand, domestic, industrial and agricultural waste in the lower reaches in Egypt is threatening water usability and availability, and human health.

Finding solutions to all these ills and to the problem of water sharing for the benefit and welfare of the people remains a major task requiring basin-wide effective, cooperative and coordinated action. Despite their many difficulties and sensitivities, the basin countries have taken some intiatives in this direction with the assistance of U.N. agencies.

The hydrometeorological survey of the catchments of lakes Victoria, Kyoga and Albert, undertaken with the assistance of the UNDP, began in 1967 and continues to the present day as inter-governmental activity under the management of a technical committee representing all the basin states. This has created a core of experts and technicians from thhe participating countries. In the project's first and second phases, basin-wide hydrometeoroligical networks have been established, together with the development of a mathematical model of the basin's upper catchment.

The governments participating in the project decided in 1988 to take action in order to establish and promote effective cooperation among the Nile riparian countries, and invited the UNDP to extend the necessary assistance for study, and to propose and establish appropriate machinery for effective cooperation among the Nile countries for harnessing the water resources of the basin. The UNDP fielded a fact-finding mission which presented its report in July 1989 (UNDP 1989). This outlined the context of regional development for the Nile, assessed the water resources potential of the Nile and the requirements of its population in the medium and long term, and suggested an action plan. The report was examined in an inter-governmental meeting and provides some useful information.

The UNDP, in cooperation with the governments of the basin countries, is now providing assistance for developing a framework for cooperation for the sustainable management and development of the region's water resources. The project aims to undertake a diagnostic study of the basin and formulate an action plan for environmentally sound management of the basin's water resources.

2. THE BASIN'S DEVELOPMENT

The Nile basin is perhaps the archetype of the usual historical pattern of international river basin development: early and significant development in the delta and lower basin and later—in this instance several thousand years later—development in the upper basin (Garetson and Hayton 1967).

Development in the Nile basin started at the early dawn of history, when the ancient Egyptians attempted the diversion of Nile waters to the low-lying fertile lands of the delta. Such developments were dictated over the years by the growing and diversified needs of the people and were influenced by the natural characteristics of the river, on the one hand, and political factors from both within and outside the basin, on the other. The beginning of this century witnessed major control works and diversions in the two downstream countries, Egypt and the Sudan, and the emergence of basin-wide plans for utilization of Nile waters. Legal frameworks developed gradually with the growing and competing needs in the basin.

2.1 The Natural Law of the Basin

The Nile systems originate in three distinct geographical and climatic zones: the Ethiopian plateau, the Equatorial lakes and the Bahr el Ghazal basin. Figure 2 provides a schematic representation of the Nile basin.

The basin extends over an area of 2.9 million km^2 and transcends nine riparian states: Burundi, Egypt, Ethiopia, Keyna, Rwanda, Tanzania, Uganda and Zaire. For almost half its length, the river runs through arid and desert lands. The total length of the river and its tributaries amounts to 37,500 km, its main lake areas total about 81,500 km^2, and its swamp areas extend over 69,720 km^2.

The rainfall over the basin which extends between 10°S and 32°N of the equator is influenced by its extent and location in the African continent, and mainly by the arrival, time duration and intensity of the inter-tropical convergence zone (ITCZ). This zone migrates annually between the tropics about six weeks behind the sun. Because of this we observe, near the equator, the passage of the ITCZ twice a year resulting in distinct peaks of rainfall. Apart from areas influenced by their coastal location or topography, there is a clear distinction of the region into zones of seasonal rainfall, all running parallel to the equator. The zones outside the equatorial region are characterized by droughts that can last for two, three or even four seasons. The pattern of annual rainfall shows some departures in the east, as the highlands in Ethiopia, Kenya and Tanzania cause relatively heavy rains. The northern tip of the basin receives winter rains varying between 100 and 50 mm.

In brief, the annual water input in the equatorial region amounts to 400 milliard m^3 (mld m^3). There are between 100 and 200 rainy days per year. The storage in Lake Victoria (69,000 km^2) amounts to 2910 mld m^3. However, what reaches the Sudanese border annually in a normal year varies between 20 and 22 mld m^3 only.

In the Ethiopian Plateau the direct rainfall amounts to 250 mld m^3 per annum. The storage in Lake Tana amounts to 28 mld m^3. It contributes to

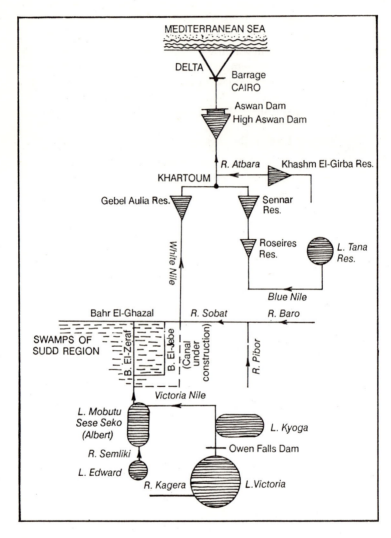

Figure 2. Schematic Representation of Nile Basin
Source: Macdonald 1993

one-seventh of the Blue Nile average discharge and constitutes the main perennial source. The total annual discharge of the Blue Nile system amounts to about 50 mld m³.

In the Sudan, the total annual rainfall amounts to about 480 mld m³, concentrated mostly in the southern region of the Sudan and varying from about 1500 mm to zero in the North, with the rainy days varying from 100 days to few days in a year, respectively. Of this rainfall, nothing reaches the Nile. Most of it, together with 50 per cent of the flow entering the Sudan from the Equatorial lakes, is lost in the Sudd region in the South.

The annual discharge that reaches Aswan in normal years amounts to 84 mld m³, from the following sources:

Blue Nile	59%
Sobat	14%
Atbara	13%
Bahr el Jebel	14%

In other words, about 85 per cent of the flow of the Nile at Aswan comes from the Ethiopian plateau and only 15 per cent from the Equatorial lakes after the losses in the Sudd region and Machar marshes. All the water contributed by the Bahr el Ghazal basin is also lost in the Sudd.

During the time of flood the contribution at Aswan is as follows:

Blue Nile	68%
Sobat	22%
Atbara	5%
Bahr el Jebel	5%

Figure 3 shows the slope of the Nile from its source to its mouth. It is important to take note of the hydrological characteristics of the different tributaries of the Nile system.

The Upper Equatorial Nile system is characterized by the natural perennial storage in the lakes, particularly in Lake Victoria, and in the Sudd region in the Sudan. The seasonal variations are considerably moderated by these natural storage systems and provide timely supplies of water.

On the other hand, the variations in the Blue Nile, Atbara and Sobat are very sharp between the wet and dry seasons. The development in such cases will require man-made storage to capture part of the flood waters for use during the low flow period. The average seasonal flow of the Blue Nile for the period 1912–82 ranged from 6200 m³/s to 125 m³/s, while in the White Nile system it ranged from 525 m³/s to 121 m³/s.

The annual variations in the flow are also very marked, particularly in the Blue Nile and Atbara basins. The annual flow of the Nile at Aswan

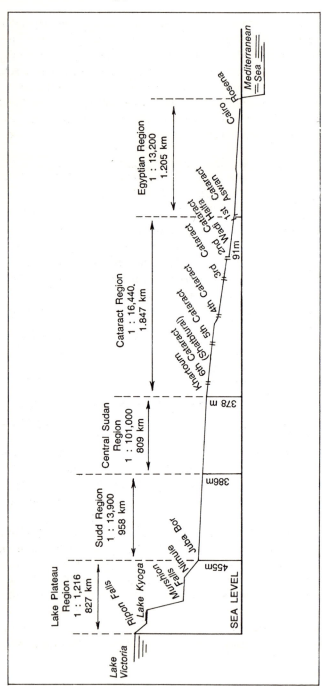

Figure 3. Slope of the Nile from Lake Victoria to the Mediterranean
Source: Hall 1991

varied from 104 mld m³ in a good year (1946) to 45 mld m³ in a poor year (1913). The average flows for the period 1870–1988 are given in the following table. This brought in the concept of over-year storage projects in the basin.

Period	Average annual flow at Aswan
1870–1954	93 mld m³
1870–1899	110 mld m³
1899–1954	83 mld m³
1954–1988	81.5 mld m³

Source: Permanent Joint Technical Commission for Nile Waters, Annual Reports.

With these peculiarities of the river and its hydrologic regime, its importance and significance for all the societies along its reaches is obvious. The particular role it plays varies. To Egypt, the Nile has been the source of life. In the Sudan, it dominates the economic and social life of the nilotics, specially in the arid and semi-arid north. This role diminishes as one moves south into the equatorial region and east into the Ethiopian highlands, where the rainfall continues for between 6 and 10 months of the year. In Ethiopia, the highlands, where the rainfall is relatively well distributed and reliable, contain 80 per cent of the population, while the lowlands, characterized by erratic and unreliable rainfall, are inhabited by nomads.

2.2 Human Interference

Various problems continue to threaten the well-being of Egypt, which is completely dependent on the Nile with its seasonal variations and annual variations—from destructive floods one year to low levels causing food shortages another year.

At the beginning of the century, modern control works were started in Egypt under the Turkish rule of Mohammad Ali Pasha. The Delta Barrage was constructed to increase the diversion of water to the fertile lands of the delta for the production of food crops and cotton. During this phase, the Egyptians penetrated south into the upper reaches of the river in search of control sites for flood protection and storage sites to augment dry season flows and meet the increasing demand for water. Turkish rule opened up the basin to European expansionism. This started with geographical and exploratory missions which were followed by military expeditions, and led to European domination of the basin in the following decades, during which time the basin became highly politicized.

The fall of the Ottoman Empire which had dominated Egypt and the Sudan and other parts of the basin marked the beginning of the era of European political influence. This had a profound impact on the socio-economic conditions of the people of the basin, the patterns of development along the basin, and on the legal framework and institutional systems. This impact continues to be felt and influences events and attitudes even in present times.

In spite of the foreign powers' subdivision of the basin into zones of influence, the hydrological unity of the basin was recognized. This was evidenced by the protocols and boundary agreements drawn up between the British, who influenced a major portion of the basin, the Italians in Ethiopia, and the Belgians who dominated the Congo Free State. The agreements include the 1891 Rome Protocol, the 1902 Agreement for the Blue Nile, and the 1906 Agreement with Congo Free State. All these agreements prohibited any construction on the tributaries of the Nile which might obstruct the flow to the Sudan and Egypt without prior consultation and agreement.

From that point, basin-wide plans started to emerge for annual storage works to control the river, in line with Egypt's interest in improving the low summer flows. There were plans to increase the Nile's yield by conserving water from the swamps and Sudd region in southern Sudan; plans for flood control works; and the over-year storage concept for the Equatorial lakes and Lake Tana in the Blue Nile basin, whereby water from high-yield years would be stored to meet shortages from low-yield years. A number of commissions were established to examine those plans and the allocation of waters between Egypt and Sudan. Based on the recommendations of the commissions, the 1929 agreement was concluded in the form of an exchange of notes (incorporating the report of the commissions) between the Egyptian Government and the Administration of the Sudan and East Africa, on behalf of the Government of Great Britain.

This agreement was motivated by the uprising of the Egyptian Army and the Sudan defence forces against the British Army in the Sudan, the '1924 Revolution', and the assassination of the Governor General of the Sudan, Sir Lee Stack, in Egypt. While the 1929 Agreement recognized the need to develop irrigation in the Sudan, it stipulated that any increase in the use of the Nile's waters as a result of such development should not infringe upon Egypt's natural and historic rights. The working arrangements based on the agreement provided for Egyptian rights over the whole of the natural flow of the river during the low flow period from January to July. Apart from small withdrawals from the natural river during this period, the Sudan had to meet its requirements from water stored at the tail end of the high flow period. So in the years that followed up to independence the development of irrigation in the Sudan was restricted to the Gezira Project, with cotton as the main crop over an area of about one million acres and a cropping intensity of 50 per cent. During the Korean war and with the sharp increase in cotton prices, further areas amounting to about 500,000 acres were developed by pumping water from the Blue and White Niles, under flood licences. Thus the 1929 Nile Water Agreement created a potential conflict between the two main downstream riparian states, the Sudan and Egypt, over the sharing of the river waters.

The 1929 Agreement also stipulated that the East African countries were not to construct any works in the Equatorial lakes without consulting Egypt and the Sudan. The British Government was to use its good offices to facilitate the establishment of over-year storage in the Equatorial lakes, linked with the conservation projects in the Sudd region to increase the Nile's yield in the interest of Egypt.

Neither the 1929 Agreement nor the project worked out to regulate the lakes for over-year storage included any reference to the riparian rights of the East African countries. However, on the eve of the independence of the East African countries in the early sixties, the British Administration of the territory presented the two downstream countries with the question of the water rights of the East African countries. This created another potential conflict between the newly independent African states and their downstream neighbours.

In the same manner, early plans formulated for the development of Lake Tana for over-year storage did not contain any clear benefits for Ethiopia or recognition of its riparian rights. This is probably the reason for the rejection by the Ethiopian authorities of all efforts to obtain concessions to use Lake Tana for over-year storage.

In 1938 the Egyptians presented the British Administration in the Sudan with the plan for the Sudd diversion canal at Jonglie. With the over-year storage in the Equatorial lakes (known as the Equatorial Nile Project) the aim of the canal was to conserve about 7.5 mld m^3 of seasonal water to improve the summer Nile flow at Aswan. The Sudan would get no benefit from these waters, but would be compensated for the disruption caused by the project which would lead to loss of livelihood for some of its people. This project was shelved.

In the forties, agreement was reached between Egypt and the British Administration in the East African territory to establish the Owen Falls Dam at the outlet of Lake Victoria. The dam could generate hydropower using the natural force of the Victoria Nile, with arrangements to enable the lake to be used for over-year storage by Egypt in the future.

Following the Second World War, a new political environment prevailed, characterized by the call for self-determination and the liberation of colonized territories. The Nile basin became an important arena of change. The events included Egypt's struggle to end the military occupation of the Suez Canal; the Sudan's struggle for independence or unity with Egypt; the Mau Mau land struggles in Kenya; and Ethiopia's struggle against the Italian occupation and influence. The Nile waters became a central issue in the politics of the area.

In 1945, the British Administration in the Sudan had established the Jonglie Investigation Team to reconsider the Egyptian Project Proposal of 1938. The team came up with a modified Equatorial Nile Project. While keeping to the original objectives of the project, it modified the storage and regulation of flow from the lakes to ensure minimum disturbances to the swamp regime in southern Sudan and to the prevailing socio-economic subsistence systems associated with it.

On the eve of the independence of the Sudan and the beginning of the Egyptian revolution in 1952, the administration in the Sudan embarked on the preparation of the case for the Sudan's share in the Nile waters. While the Sudan had a genuine case concerning the Nile water, the political motives cannot be overlooked, in view of the phase of conflict that followed between the two sister countries. The period 1954–58 witnessed further developments: the emergence of plans to extend the irrigated area in the Sudan and build the Roseries dam on the Blue Nile, which required an increase in the Sudan's share of the Nile waters; and the Egyptian plan for over-year storage at the Aswan High Dam, with its reservoir extending 150 km into the Sudan and completely submerging the old town of Halfa and all the villages in the area, affecting a population of about 50,000. At the same time, in 1954, the report on the Nile Valley Plan was being written by H. A. Morrice and W. N. Allan, advisors to the Sudan Government.

The Nile Valley Plan was a purely hydraulic plan, designed to improve the distribution and utilization of surface water, mainly by means of dams to store that water before it was lost to the sea. As the authors of the plan stated, 'we wish to emphasize that our investigations have been confined to the hydraulic aspect of the matter. We feel confident that the Nile Valley Plan we advocate is in essence economically sound, but we have made no attempt at an economic analysis. We fully realize that such an analysis must eventually be made for each component project, but we believe that the first step must be to draw up an outline plan based on the hydrological characteristics of the Nile Valley.'

The Nile Valley Plan was primarily an attempt to control the Nile and its tributaries, in order to assure the largest amount of water for irrigation, particularly for the Sudan and Egypt, and the full development of the hydroelectric potential of the Nile and its tributaries. Due to a lack of information, the plan had to assume the amounts to be abstracted by Ethiopia and the East African territories. The plan thus lacked economic and environmental dimensions, and all these years it has remained in the archives. However, it is rated as an important scientific contribution still worthy of examination, even in the context of the new environmental complexities.

The emergence of the High Aswan Dam Plan led to further complications, transcending the boundaries of the basin and moving into international politics. Following the Western countries' and the World Bank's decision to withdraw the initial support they had given the project, the Egyptian Government approached the Eastern Bloc (the erstwhile USSR) for technical and financial support and, in retaliation, nationalized the Suez Canal. This was followed by the tripartite aggression against Egypt in 1956 which put the High Aswan Dam Plan into the international spotlight. A war was waged worldwide against the project's environmental and socio-economic impacts.

Within the basin, the period 1954–58 witnessed political conflicts between the Sudan and Egypt over the High Aswan Dam Plan and the sharing of water. Negotiations came to an impasse when the Sudan declared unilateral non-adherence to the arrangements of the 1929 Nile Water Agreement and there was increased pressure for new arrangements to increase the Sudan's share.

After the military takeover in the Sudan in November 1958, the Sudan and Egypt concluded a new legal and technical agreement in November 1959 to replace the 1929 agreement. Egypt was alowed to go ahead with plans to establish the High Aswan Dam for over-year storage, arresting and controlling the full discharge of the river at Aswan, amounting to 84 mld m³. The net total benefit of the project, 22 mld m³, would be divided, giving 7.5 mld m³ to the Sudan, and sharing equally the 10 mld m³ which would be lost by evaporation at the Aswan reservoir. This new water allocation increased the Sudan's share to 18.5 mld m³ and Egypt's share to 55.5 mld m³. The High Aswan Dam Project, with its over-year storage, has protected Egypt against floods as well as drought years and enabled full control of the river in Egypt's own territory. To a great extent, the dam mitigated the conflicts between the Sudan and Egypt caused basically by the restrictions imposed by the 1929 Agreement prohibiting withdrawal by the Sudan during December to July and increasing Egypt's share of water taken. The new agreement recognized other riparian rights and stipulated that the agreed amount should be shared equally.

The two countries also agreed to establish a Permanent Joint Technical Commission to undertake, on behalf of the two Governments, the control of the river, and to carry out studies to increase water yield to meet the future demands of the two countries.

The 1959 Nile water agreement between Egypt and the Sudan brought a number of reactions from the other riparian states. Ethiopia stressed its legitimate rights to the waters of the rivers originating from its plateau. With the assistance of the United States Bureau of Reclamation, it initiated studies to identify power and irrigation projects within the Blue Nile, Atbara and the Baro of the Sobat system. This identification was followed by an assessment of water resources and the irrigation and power potential. A Water Resources Commission was established to undertake water resources management and development.

On the eve of the independence of the East African countries and following the 1959 Nile Water Agreement between Egypt and the Sudan, the British Administration of the territory brought to the attention of the governments of these two downstream countries, the East African Countries' claims to water rights and the need to make new arrangements to supersede the 1929 Nile Water Agreement.

It was agreed that informal technical talks would be initiated between the Permanent Joint Technical Commission, representing the two downstream countries, and the coordinating Nile Water Committee (established for the purpose) representing the East African countries— Kenya, Tanganyika and Uganda. During the talks it became apparent that the Administration had no ready plans to indicate and substantiate the water requirements of those latter countries. It was therefore agreed that joint studies would be initiated in the catchments of Lakes Victoria, Kyoga and Albert to determine the water balance of the lake area, obtain the required data and information, and identify the necessary storage work to meet the future demands of the riparian states. Other countries were invited to join in this basin-wide cooperation, including Burundi, Rwanda, Zaire and Ethiopia. All the countries agreed to join in this effort, with the exception of Ethiopia, which opted to join as an observer.

With the assistance of the United Nations Development Programme (UNDP) and with the World Meteorolical Organization (WMO) as executing agency, the Hydromet Survey of the Equatorial Lakes was launched in 1967. A Technical Committee was established with representatives from all participating countries, with Ethiopia as an observer to oversee the project on behalf of the governments of the basin. Counterpart staff and counterpart funds were supplied and the project headquarters was established in Entebbe, Uganda. This was one of the most successful institutions of the basin, being the first forum for cooperation, despite the fact that in terms of area it extended only to the lake catchments of the equatorial region.

After its successful completion, the project was extended to a second phase, with further assistance from the UNDP for the formulation of a mathematical model representing the Upper Nile at the Equatorial Lakes. Efforts to extend the model to include other reaches of the river could not be concluded. The Hydromet Project, now fully administered by the Technical Committee and financed by the participating governments, continues to collect hydrometeorological data and carry out analysis. The plans made since the late seventies to develop the Technical Committee into a basin authority and widen the scope of its functions seem to have come up against political suspicions that have accumulated over the years.

In 1976, the two downstream countries started to construct the first phase of the Jonglie Canal as the first conservation project to increase the Nile's yield. The project planning and implementation came at a time of heightened environmental awareness. Like the High Aswan Project, the Jonglie Canal received very wide attention within and outside the basin and became highly politicized. The first phase of the project concept departed very much from the original Equatorial Nile Project. It is confined to a diversion of 20 mm^3/day from the Sudd area, without the need for storage in the lakes, with a water benefit of about 5 mld m^3 shared equally between the two countries. This is compared to the original project which required storage at the lake and the diversion of 55 mm^3/day, with a water benefit to Egypt of 7.5 mld m^3. Unfortunately, construction work was suspended halfway due to circumstances in the Sudan.

3. CHALLENGES AND OPPORTUNITIES

While all the countries of the basin have recognized the need for and expressed a commitment to cooperation within the basin for its integrated development and management, unilateral actions continue to prevail. This includes the preparation of national master plans to evaluate the water and land potential for irrigation and hydro-power development within each country of the basin, with no coordination or consultation between them.

The two downstream countries, Egypt and the Sudan, base their water development plans on the shares stipulated in the 1959 Nile water agreement and on future conservation projects in the Sudd area to increase the yield of the river. However, the two countries recognize the riparian rights of upstream countries and are aware that such rights will have to be taken into account from the point of view of present shares and any future increments that can become available from the conservation projects. On the other hand, the upstream countries, particularly Ethiopia, while fully subscribing to the principles of integrated development of the basin, stress their right to the use of the waters of the Nile on the basis of equitable sharing. The upstream countries are preparing independent plans for irrigation and hydro-power development and are assessing their water needs accordingly.

The Nile basin continues to have vast potential to meet the needs and requirements of its societies, under conditions of cooperative and coordinated basin-wide action. The environmental stability and integrity of the basin are vital for its sustainable development.

The countries of the basin are among the least developed countries in the world, with agriculture as the primary sector. Over 80 per cent of the population in almost all the countries of the basin are engaged in agricultural production, particularly in the rainfed and livestock subsector in the upper reaches of the basin and in the irrigated subsectors in the arid and semi-arid lower reaches.

The total population of the countries of the basin stands at 246 million (over 75 per cent live in the basin proper), and is growing at an average rate of 2.5–3.0 per cent per annum. The majority of the inhabitants live in absolute poverty, lacking basic needs in terms of food, water, energy and shelter. They exert formidable pressure on the fragile environment of the basin, and future demographic changes will further increase this pressure, leading to grave environmental consequences unless concerted coordinated action is taken at the basin level.

In its upper reaches the basin is witnessing severe erosion and top soil loss, with complex sedimentation and morphological changes in its lower reaches. In the Ethiopian highlands, the rate of erosion is estimated to be 100–300 thousand million tons of top soil per year, which is equivalent to 120,000–210,000 hectares of land with one metre depth of soil. According to recent estimates, the destruction of forests has taken place at the rate of 200,000 hectares per year. Water-induced erosion and associated biological soil degradation have considerably affected the long-term productivity of the land. Intense storms, barren land slopes, removal of forest and vegetation cover, and land misuse through the shortening of fallow periods are the main factors responsible for the severe soil erosion. The Ethiopian Highland Reclamation Study carried out by the Food and Agriculture Organization of the United Nations (FAO) has provided a valuable base of information on this problem. There is a great need to extend the study to the downstream reaches of the Nile tributaries originating from the Ethiopian highlands, which are affected by morphological and sedimentation problems.

The major findings of the highland study revealed the need to promote a conservation-based development strategy. The implementation of such a strategy is beyond the capability of one country. Ethiopia's conservation efforts during the last decade have been localized. What is required is a massive and total commitment from all the Nile basin countries, those that are directly affected and as well as those that are indirectly affected.

Food production in the basin is mainly concentrated in the rainfed areas and in the semi-arid savannah belt. The majority of the people of the basin are engaged in rainfed agriculture. The fertility of these areas has been affected by many factors, including loss of top soil by erosion, depletion of nutrients due to heavy rain, lack of drainage systems, and deficient soil/water-management techniques. However, the major cause of soil degradation and its declining productivity is the increasing population pressure on the land. In the semi-arid zone, desertification is capturing significant areas. The persistent drought that was witnessed in the last two decades reached its highest level in 1984, contributing to huge food gaps. Many pockets in the basin were hit by famine, affecting more than 15 million inhabitants. In such a catastrophic situation, further destruction of the basin's resource base takes place. According to the World Bank's *World Development Report 1989*, food production in Ethiopia dropped by 30 per cent from its 1961 level while the population had increased at an annual rate of 2.5 per cent. To give another example, the level of food production in the Sudan was the same in 1987 as in 1961, with marked fluctuations from year to year, depending on the rains.

The years 1980–85 witnessed a sharp decline in food production, which had a marked influence on the collapse of the environmental and socio-economic systems in the semi-arid zone and on the huge demographic changes associated with it. The mass migration from rural to urban centres witnessed in parts of the basin not only weakened the production base, but also brought dangerous environmental consequences in the urban areas, straining urban facilities to the point of breakdown, with associated health hazards.

Irrigated agriculture also continues to be a major activity in the basin. It is the largest user of the basin's surface water resources and a primary sector of economic growth in the downstream countries, Egypt and the Sudan. In Egypt, irrigated agriculture is the dominant sector. Over 4 million ha are under irrigation, and there are plans to expand over an area of another 1 million ha. In the Sudan, the irrigation subsector contributes 65 per cent of the GNP and extends over an area of 1.5 million ha, with plans to expand over an area of another 1.5 million ha.

The role of the irrigation subsector diminishes as one moves to the upper reaches of the river basin. Present irrigation in the Upper White Nile riparian areas is very limited, through there are plans for future expansion over an area of 130,000 ha in Uganda; 200,000 ha in the United Republic of Tanzania; and 57,000 ha in Kenya. In Ethiopia, too, much of the irrigation is practised by traditional farmers, and is not significant. According to a 1984 FAO study, the potential identified in the Blue Nile basin includes 100,000 ha of perennial irrigation, requiring storage, and 165,000 ha of small-scale seasonal irrigation. The other riparian countries, Burundi, Rwanda and Zaire, have no potential for irrigation in the basin and depend almost completely on rainfed agriculture (UNDP fact-finding mission report).

The expansion of irrigated agriculture in the years ahead will require basin-wide cooperation in the management of water resources to meet increasing demands and to face the environmental consequences associated with them. The downstream states have vast experience in this sector, which, through cooperation, could be put at the disposal of the upstream countries. Meeting the increasing demands for irrigation will require careful planning and development of control works and conservation techniques in the different reaches of the river in terms of storage and swamp reclamation works to increase the river's yield. These pose major challenges, requiring basin-wide cooperation for sound environmental management of the basin's water resources.

The hydro-power potential of the Nile basin also offers vast opportunities, particularly in the upper reaches of the Blue Nile and the White Nile, and the main Nile in northern Sudan, and offers great scope for basin-wide networks reaching arid regions. Along the river, there are no apparent conflicts between power and irrigation demands. On the contrary, the development of the power potential in the upper reaches of the basin will tend to improve the dry season flows of the river to meet irrigation demands and will open up opportunities for basin-wide cooperation in this respect.

Though it is not certain, there is a fear that the changing trends in annual precipitation yields and patterns which have been witnessed in the Nile basin could be attributed to climate change. Increasing greenhouse gas concentrations could influence the discharge regime of the basin. This is another challenge that requires technical and scientific capabilities, with concerted cooperation within the basin. Persistent drought and the associated environmental degradation continue to hit many parts of the basin and are matters of great concern for the future integrity of the basin's resource base.

4. INSTITUTIONAL FRAMEWORK

Within international river basins, basin-wide cooperation essentially needs to be built upon a good foundation of national institutions. Therefore, it is essential to review the present institutional systems in the different countries of the Nile basin that are engaged in the management and development of water resources. We can then review intergovernmental activities towards basin-wide cooperation.

The Egyptian Department of Irrigation, recently renamed the Ministry of Public Works and Water Resources, is one of the oldest institutions in the Nile basin and is responsible for the development and operation of irrigation systems. Its functions have widened over the years to include Nile control and hydraulic research. It was responsible for the hydrological network of the Nile in the Sudan and Egypt, which has now been taken over by the Permanent Joint Technical Commission on Nile Waters.

An important institutional development in Egypt is the creation of the Water Research Centre (WRC), which has eleven institutes addressing research into and management of water resources. Five of these institutes bear directly on the Nile waters: the Nile Research Institute; the Water Distribution and Irrigation Systems Research Institute; the Drainage Research Institute; the Water Resources Development Research Institute; and the Hydraulic and Sediment Research Institute. The WRC has provided the technical assistance for some extremely interesting research programmes and is developing promising scientific capabilities.

The Department of Irrigation in the Sudan was created in 1925 to operate and maintain the Gezira Irrigation Project. After independence in 1956, the department was upgraded to become the Ministry of Irrigation and Hydro-power, responsible for the development and management of irrigation and Nile control and the development of hydro-power in the Sudan. The hydro-power department's functions were later transferred to the Ministry of Energy and Mineral Resources, but the functions are carried out in full coordination with the Ministry of Irrigation. The latter now has a department of water resources and a hydraulic research station, established during the mid-seventies with the assistance of the United Nations Educational, Scientific and Cultural Organization (UNESCO).

In Ethiopia, the water resources department of the Ministry of Public Works and Communications was established in 1958 to handle multipurpose investigations of the Blue Nile basin. In 1971, the National Water Resources Commission was established, with broad responsibilities and functions in the field of water resources. Due to financial constraints and lack of trained manpower, the Commission could not fully exercise its mandate. In 1977, the Valley Agricultural Development Authority was formed, with jurisdiction over the country's water resources. At the same time, the Ministry of Agriculture was empowered to investigate, use, control, protect and administer the water resources of Ethiopia for irrigated agriculture and other purposes. In a further attempt to improve coordination and avoid duplication, the National Water Resources Commission was restructured to absorb the function of the Valley Agricultural Development Authority. The Commission encompasses four authorities: the Water Resources Development Authority; the Water Supply and Sewage Authority; the Ethiopian Water Works Construction Authority; and the National Meteorological Services Agency. The Commission was given the functions of conducting studies to utilize, administer, regulate, protect and locate inland water, and of supervising government water policies and plans. Following the 1984 drought, and in light of associated problems that were beyond the technical and financial abilities of the Commission, a new institution was created in 1987: the Ethiopian Valleys Development Studies Authority. The main functions of this new authority are to prepare country-wide and basin-wide master plans for the use of water and related resources, and to investigate water projects to assess their feasibility.

In the East African countries, Kenya, Uganda and the United Republic of Tanzania, water resources fall under the Ministries of Water and Mineral Resources, which are mainly concerned with operational hydrological activities. In Tanzania, considerable emphasis has been placed on the importance of irrigation. In 1975, a water development and irrigation division was established within the Ministry of Agriculture, with the help of the Sudan Ministry of Irrigation. The programme has met with many difficulties, and performance has not been up to the Government's expectations. The Institutional Support Project sponsored by the FAO aims to correct the constraints and build up adequate capacity in the irrigation division.

In Burundi, Rwanda and Zaire the style and structure of the institutional systems are different from those in other parts of the basin. In Burundi, the hydro-meteorological responsibilities rest with the Geographic Institute of Burundi which comprises four departments: the Department of Synoptic Meteorology; the Department of Hydrology; the Department of Agroclimatology; and the Department of Instrumentation.

In Rwanda, several government units deal with water. These include the Directorate of Meteorology in the Ministry of Transport and Communication, the Directorate of Water in the Ministry of Public Works, Energy and Water; and the Institute of Agronomic Science.

While no activities exist in the basin within Zaire, the institutional framework comprises the Ministry of Land Tenure, Environment and Conservation of Nature, which is responsible for hydrological activities in the country, and a parastatal body, REGIDESO, which is responsible for water allocation in the country.

At the subregional level a number of experiences can be quoted. The Permanent Joint Technical Commission for Nile Waters, established between Egypt and Sudan in 1960 following the 1959 Nile Water Agreement, is a good example of subregional cooperation in the basin. The functions of the Commission include supervising the control and gauging of the river between the two countries and allocating shares; undertaking studies to increase the Nile's yield and meet future demands; and promoting basin-wide cooperation. However, the focus of the Commission is purely engineering, lacking the multidisciplinary approach required for integrated water resources management and development.

The Hydromet Project of the Equatorial lakes, established with assistance from the UNDP, is another example of subregional cooperation in the basin. Despite the limited scope of its functions, the Technical Committee that has overseen the project for the last 25 years has currently taken initiatives, with the assistance of the UNDP, to further the objectives of this cooperation to embrace the whole basin.

At the initiative of UNDP, the first meeting of the ministers from the Nile countries was held in Bangkok in January 1986. It was attended by ministers involved in water resources management and development from Egypt, the Sudan, Uganda, Tanzania and Zaire, and by the ambassador of Ethiopia to France and high-level officials from Burundi and Rwanda. The ministers, referring to the experience of the Mekong, decided to take action to promote and establish effective cooperation among the Nile riparian countries at the earliest possible opportunity. They invited the UNDP to provide the necessary assistance for studying, proposing and establishing appropriate machinery for effective cooperation among the Nile countries to harness the river's water resources.

The UNDP responded by providing financial assistance to support a fact-finding mission and to organize a second meeting of the ministers in Addis Ababa in January 1989 to review the appropriate mechanism for regional cooperation. The terms of reference of the mission were finalized under the auspices of the Technical Committee of the Hydromet Project of the Equatorial Lakes in February 1989. The terms of reference aimed, in particular, at identifying the interests of the countries of the basin in joint regional development, controlling and managing the basin's water resources; specifying concrete regional development activities, schemes and programmes; assessing the status of water resources and of development activities and programmes carried out to date; and evaluating the national capabilities. The mission would also try and estimate the extent of external assistance needed on a regional and national basis.

The report of the fact-finding mission outlined the context of regional development conceived for the Nile countries, presented an assessment of the water resources potential and demands in the medium and long term and suggested an action plan to control water resources. The report was to serve as basic material for the third meeting of the ministerial committee, later in 1989.

The Nile basin countries are now engaged, with the assistance of the United Nations Environment Programme (UNEP), in preparing a diagnostic study of the basin and preparation of the Nile Action Plan for environmentally sound management of the basin water resources. This it is hoped will lead to basin-wide cooperation.

5. LESSONS FROM THE PAST

It is now widely recognized that 'The last decade of the twentieth century is a time of great promise, great risk and great complexity. Events are accelerating in several fronts simultaneously—economic, ecological and political—and are forcing profound changes in the relationships among people, nations and governments.' (MacNiell 1991).

It is in this context that we take stock of the situation in the Nile basin in relation to the risks, compexities and past conflicts.

Water conflicts are now routine matters of life. They happen between regions in one country or between countries sharing a river basin, and could be over the quantity or the quality of water.

The Nile basin is similar to the basins of many international rivers which have been potential sources of conflict. Such conflicts have surfaced many times in history, have developed over time and taken different forms, and have been dealt with as they have arisen.

The main competition for Nile Waters is between the two downstream countries of the basin, the Sudan and Egypt. This is dictated by their locations in the arid and semi-arid zones of the basin and the hydrological characteristics of the rivers in their zones.

The 1929 Nile water agreement, between Great Britain and Egypt, gave Egypt exclusive rights over the dry season flow of the river, from January to December. As far as Egypt and the Sudan are concerned this is history, as it was superseded by the 1959 Nile waters agreement. But it still binds the East African countries not to construct any works or modify the flow without consultation and agreement with the two downstream countries.

It is important to hark back to the 1929 Nile agreement to learn some lessons which could be useful in guiding future action and cooperation.

From a political point of view and interpretation, the agreement infringes on national sovereignty, and could be said to have made unfair allocations. But from the technical viewpoint, whether the agreement existed or not, the Sudan would not be able to expand its use of water from the Blue Nile and Atbara during the period January to June without reverting to annual storage of part of the flood waters to use during the dry season. Both rivers are seasonal and dry out completely during January–June, apart from a small supply from Lake Tana and a little unreliable rain in the basin of the Blue Nile. The 1929 agreement did not bar the Sudan from expanding its use of Nile waters by providing storage works in both rivers. There are no convincing reasons why the British administration did not go ahead with the implementation of Rosieres and Khashm El Girba dams on the Blue Nile and Atbara rivers respectively. It was definitely not because of the 1929 agreement, as the Sudan could have expanded its use without infringing on Egypt's rights under the agreement. The interpretation of such a situation could be that it was a deliberate policy of the colonizer to keep development slow in the Sudan. The 1929 agreement may have unfairly restricted the Sudan's use of the White Nile and Main Nile, but the potential for irrigated agriculture mostly lay in the Blue Nile basin.

On the eve of the independence of the Sudan in 1955, the conflict over the sharing of Nile waters began to gain importance and the matter became a hot political issue. Interim arrangements were made by providing extra storage in the Jebel Aulia reservoir in favour of Egypt to allow the Sudan to increase its abstraction from the Blue Nile during the restricted period to meet water needs for the Managil irrigation extension.

The conflict between the two countries was finally resolved when the concept of over-year storage at the Aswan High Dam came into being in the 1959 Nile agreement. The restriction imposed by the 1929 agreement was lifted, and it was possible to store 32 mld m^3 of water that would have otherwise found its way to the sea (flood waters of the Blue Nile and Atbara rivers). The net increase in available water of 22 mld m^3 (10 mld m^3 evaporation loss) was shared between the two countries, with 14.5 mld m^3 to the Sudan (raising its share to 18.5 mld m^3) and 7.5 mld m^3 to Egypt (raising its share to 55.5 mld m^3). However, this project aroused a good deal of opposition within Egypt, within the Sudan and in international circles.

Some circles in Egypt are concerned by the degradation of the Nile Channel and loss of soil fertilty caused by arresting the silt behind the dam. Some other objections were politically motivated against the government at that time. In the Sudan the opposition took other directions, particularly from those directly affected by the submergence of their homes. Many of them had to resettle over a thousand miles away in a completely different, though economically superior, environment. Others were concerned that the water sharing was still unfair.

Outside the two countries the reactions were of two types. First, the upstream riparian states expressed anger at the fact that the two downstream countries had divided all the water that reached Aswan between themselves, neglecting their neighbours' legitimate rights on these waters. This created an atmosphere of passive conflict which has prevailed to the present times. On the other hand, international circles were partly concerned about the environmental implications of the dam and partly about the threats to archaeological treasures, but mainly, as it proved, their reaction was a political one against the Egyptian revolution.

Over the years the High Aswan project proved that it was the only option available at that time to manage the conflict between the two main downstream countries, Egypt and the Sudan. It has proved its economic viability to Egypt and has been as an effective shield against the floods and droughts that have occurred since its implementation. The environmental impacts are manageable and within reach of control and mitigation (Biswas 1992).

Resolving the upstream-downstream conflict is vital for a smooth reconciliation process over the whole basin. The 1959 Nile agreement is an important tool in this respect. It is a bilateral agreement between the two downstream countries, not in any way binding the other riparians, but recognizes clearly the rights of other riparians. The Permanent Joint Technical Commission has been set up for control of the river and cooperation to increase its yield to meet the future demands of the two countries. The Commission has laid the foundation for future basin-wide cooperation.

The 1959 agreement between Egypt and Sudan no doubt created a rift in the Nile riparian relations, particularly with Ethiopia; but on the other hand it created an opportunity to undertake informal technical talks over the requirements of the East African countries which led to cooperation on the Hydromet Survey of the Catchments of the Equatorial Lakes which set the stage for wider cooperation.

Before going further it is important to ask the question 'is the rift between the riparians created by the 1959 agreement justified?' The answer is, politically, Yes. But from the technical or legal point of view the answer would be No.

Technically, if these two countries had been upstream and had entered into such an agreement this would have affected the flow downstream and inflicted harm on the downstream countries. As they are both downstream countries this problem does not arise. Further, this agreement is bilateral and not binding in any way on the others. Ever since the agreement came into force in 1959, there is no evidence that it stood in the way of any water requirements of the upstream countries or inflicted any harm on them. There were no serious incidents of conflict over these matters that can be cited. Abstractions of water by the upstream countries from the Blue Nile and the White Nile have been extremely small and have not affected the flow downstream.

Nevertheless, the 1959 bilateral agreement between the downstream countries has been responsible for retarding cooperation and encouraging unilateral action, and in certain instances even hindering exchange of information. At times of political confrontations that occurred between neighbours for reasons that had nothing to do with water, water became an issue. Rainfall data or a river stage became a state secret. Those were the attitudes that developed in the cold war era. The 1959 agreement should be taken in its proper context and be promoted rather than making its abolition a condition for cooperation. This more positive spirit has now started to emerge, and the UNEP and UNDP have taken initiatives in this direction.

In the long term some upstream/downstream problems may develop over the waters of the Blue Nile, but these are not unsurmountable. The elements that unite are far stronger and cooperation in the end will prevail. The White Nile is not likely to be a source of conflict.

Another potential area of conflict is the Sudd region in southern Sudan where conservation projects have been proposed. Many lessons can be learnt from the first conservation project on which work has begun, the Jonglie canal project. This led to conflict within Sudan, which needs to be viewed in its proper context. The Southern question is a deeply rooted political problem in Sudan, outside the scope of this paper. Water is just one of the elements in this conflict. What can be asserted here is that development of the water resources of the Sudd region is a key element in the development of the southern region of Sudan and the welfare of its societies.

Random allegations were made that the Jonglie canal project would transfer water to the Arab north and to Egypt, and that a million Egyptians were to be housed in southern Sudan. These statements were politically motivated and aimed to deepen the conflict in southern Sudan and create rivalries between the riparian states. Thus, the water conservation projects were used as an excuse for conflict, while on the contrary they could be an important element in making peace in Sudan and in furthering the interests of the Nile basin as a whole. The Jonglie canal project, though not yet completed because of unrest in southern Sudan, is one of the most carefully planned projects, particularly from the environmental aspect. When it is completed and becomes operational, it will provide important data on conservation technology and indicate directions to be taken in the future (Mageed 1985).

Statements and speculations such as 'The Nile is a war waiting to start' (MacNeill 1991) are not based on scientific evidence nor on the political atmosphere that should prevail in the basin in the post-cold-war era. It must be hoped that in the future the question of waters can be resolved without being complicated by other conflicts in the area.

6. ENVIRONMENTAL RISKS

As has been mentioned earlier, as a result of increasing and uncontrolled population pressures, the integrity of the resource base of the Nile basin is at considerable risk in the future. Many parts of the basin have already witnessed environmental degradation. The soil erosion in the Ethiopian highlands and its implications downstream, soil degradation, and spread of desertification in the semi-arid and arid zones, all have grave consequences. The production base in pastures, rainfed and irrigated lands is declining in many parts of the basin and their sustainability is at risk. The drought cycles and the threat of hunger in many pockets are creating panic among the basin societies, leading to demographic changes and mass rural migration to already strained urban centres. Global environment problems, including greenhouse emissions, global warming and sea level rise are future threats to the basin.

Dealing with these challenges and risks is beyond the capability of any single basin-country and will require basin-wide cooperation rather than confrontations.

The Promise

The hydrological unity of the Nile and the economic integrity of the basin, despite all these challenges and risks, offer vast opportunities in terms of food, energy and material production. Many new and emerging technologies offer great promise in this respect for increasing food production, raising industrial output, conserving the natural resources and managing the environment.

Water and water-related issues are intertwined with all these matters and have a pivotal role to play. This promise cannot be realized without developing new attitudes which break free of the old and obsolete hydropolitics. The basin water resources in terms of rainfall, surface run-off and groundwater are much more than the 84 mld m³ captured at Aswan. Even the 84 mld m³ can be used more effectively than they are today.

The promise cannot be realized until we all see the pivotal role of water resources in this broad context and mobilize the resources of the basin as a whole. A total framework is required for basin-wide cooperation in water, agriculture and food production, communication and transport, trade, population control and environment. A model similar to that of the Southern African Development Conference, adapted to suit the conditions of the Nile basin, could be considered as a basis for future cooperation. The concept of sharing the waters of the basin needs to be broadened and taken in the context of sustainable development and maintenance of the basin environment and integrity, to meet present and future demands.

REFERENCES

Allan, W. N. and H. A. Morrice (1954). *The Nile Valley Plan Report of the Sudan Government*, vol. 1.

Bashir, M. O. (1986). *The Nile Valley Countries: Continuity and Change*. Sudanese Library Series, Vol. I, No. 12, Institute of African and Asian Studies, University of Khartoum, pp. 3–5.

Biswas, A. K. (1992). 'The Aswan High Dam Revisited', *Ecodecision*, No. 3, pp. 67–69.

FAO (1991). 'Agricultural Water Use—Assessment of Progress in the Implementation of the Mar del Plata Action Plan', Report of the Regional Assessment Mission, 1991, pp. 2–16.

Garetson, A. H. and R. D. Hayton (1967). *The Law of International Rivers*. S. Book No. 320, New York, Nile Basin Integrated Development.

Hall, John (1911). *Contribution to the Geography of Egypt*.

Lodwig, Emile (1937). *The Nile* (New York).

MacDonald, Sir M., and Partners Ltd. (1983). Sudan Government, Jonglei Canal Project, Appraisal Study: Final Report, July 1983.

MacNeill, Jim (ed.) (1991). *Beyond Interdependence*. New York, Oxford University Press, pp. 3–56.

Mageed, Y. A. (1985). 'Integrated river basin development: the challenges to the Nile basin countries' in *Strategies for River Basin Management*, edited by Jan Lundquist, Lohm & M. Falkenmark, D. Reidel Pub. Company, pp. 151–60.

———— (1985). 'Conservation Project of the Nile: The Jonglie Canal' in *Large Scale Water Transfer* edited by G. Golubev and A. K. Biswas, Tycooly Pub., Oxford, England, pp. 85–101.

Rodney, Hewett, (1986). 'Assessment of Irrigation Potential in the Ethiopian Highlands', Proceedings of the National Workshop on Food Strategies for Ethiopia, Alemay University of Agriculture.

UNDP (1989). *Nile Basin Integrated Development* AF/86/003–RAB/86/014.

OFFICIAL DOCUMENTS

– Protocol between Britain and Italy, delimiting spheres of influence in East Africa, 15 April 1891. Art. 3, 83 Brit. and Foreign State papers 21.

– Treaties between Britain & Italy and Ethiopia, relative to frontier between the Sudan, Ethiopia, and Eriteria, 15 May 1902. Art.3.Cmd No.13070 (T.S. 16 of 1902), 23 Hertslet, Commercial Treaties 2.

– Agreement signed in Brussel, 9 May 1906. Art.Cmd No. 2920 (T.S. 4 of 1906), 24 Hertslet, Commercial Treaties 344.

– Exchange of Notes, regulating the use of Nile waters for irrigation, between Great Britain and Egypt, 7 May 1929. Cmd No. 3348 (T.S. 17 of 1929), 21 Mortens H. R. G. (3e Ser) 97.

– Agreement of full utilization of Nile Waters between Egypt and Sudan, 8 Nov. 1959. Text 15, Revue Egyptian de Droit Internationale 321–29 (1959).

7 / Management of International Water Resources: Some Recent Developments

ASIT K. BISWAS

INTRODUCTION

Ever since the time of Aristotle, concern has continued to be expressed on whether enough natural resources would be available for human consumption for future generations. With a steadily increasing global population, and mankind's eternal quest for a higher standard of living for all the world's citizens, there is no doubt that demands on natural resources will continue to increase as well. Even if it was possible by some miracle to stabilize world population at the present level, resource requirements would still continue to increase for a considerable period of time as more and more people achieve a better quality of life. Water is a good example of a resource for which demand is increasing continuously.

There is no question that it is going to be an increasingly complex task to provide an adequate quantity and quality of water for various human needs. Difficult as it is to institute more rational and efficient management policies and practices for water sources that are contained wholly within the geographical boundaries of individual sovereign states, for a variety of interrelated technical, economical, social, institutional and political reasons, the problem is highly intensified when management and development processes for water sources that are shared by two or more countries are considered. This is evident if one analyses the problems that have already arisen on the development and management of international water bodies—rivers, lakes and aquifers—in various parts of the world. It is especially true for arid and semi-arid regions of the world, where the vast majority of people of developing countries live, and where population growth rates are generally the highest at present.

WATER CRISIS AND INTERNATIONAL WATER SYSTEMS

Most of the countries located in arid and semi-arid regions are already facing a water crisis, though the intensity and extent of that crisis could vary from one country to another, and with time. If the current trends continue, the water crisis will become widespread and more pervasive in nearly all arid and semi-arid countries by the early part of the 21st century. For example, current projections indicate that, because of supply constraints, by the year 2000 only three countries in the Middle East—Turkey, Iran and the Sudan—may have a per capita water consumption level above the currently accepted minimum.

There are many interrelated reasons which contribute to this crisis, and only the four major ones will be discussed herein.

The first is the global population which continues to increase steadily, with attendant implications for water quantity and quality. Estimates indicate that the current world population is likely to double to 10.64 billion by the year 2050. Developing countries, which are all in tropical and semi-tropical regions, will account for some 87 per cent of this population, or 9.29 billion.

While there is no one-to-one relationship between population growth and higher water requirements, it is evident that with a substantial increase in world population, total water requirements for various uses will increase as well. Furthermore, past experiences indicate that as the standard of living increases, so does the per capita water requirement. These two factors are expected to account for a nearly tenfold increase in the total global water use in the present century (Biswas 1992a), as shown in Figure 1. This means that if the developing countries' current poverty

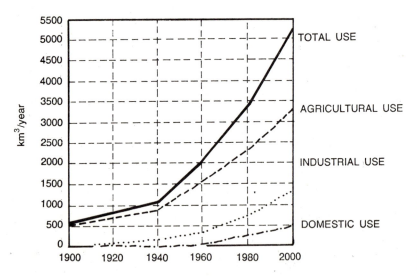

Figure 1. Increase in global water use, 1900–2000

alleviation programmes succeed, their rates of growth in water require-ment are likely to accelerate even further. This is a fact that has thus far been generally overlooked by national planners and decision-makers as well as international organizations.

Second, from an economic viewpoint, the amount of fresh water avail-able to any country on a long-term basis is limited. Since in arid and semi-arid countries nearly all the easily available sources of water have now been developed or are in the process of development, the unit costs of future projects in real terms can only be higher. For example, a recent review of domestic water supply projects supported by the World Bank indicates that the cost per cubic metre of water for the next generation of projects is often two to three times higher than that for the present generation. This is an important consideration, since most developing countries are now saddled with very high levels of debt burdens, and the amount of new investments available, both internally and externally, is limited. In addition, the demands and competition for whatever funds are available are intense. These factors, both individually and collectively, are bound to have a serious effect on the next as well as later generations of water projects, adversely in most cases.

Third, as human activities increase, more and more waste products are contaminating the available sources of surface water and groundwater. Among the major contaminants are untreated or partially treated sewage, agricultural chemicals and industrial effluents. These contaminants are seriously affecting the quality of water, especially for domestic uses. Already many sources of water near urban centres of developing coun-tries have been severely contaminated, thus impairing their potential safe use in a cost-effective manner. This in effect means that serious water quality deterioration could be considered to be equivalent to reduction in the quantity of water available for various uses in the future.

The fourth major factor is the increasing delays that are likely to be witnessed in the coming decades in implementing new water projects. In addition to escalating project costs, lack of investment funds, and increasing technical complexities of new development projects, other factors like social and environmental implications of the water develop-ment projects, which are becoming increasingly significant, are likely to delay project intiation time, certainly more than what has been witnessed in earlier decades. At least in the short to medium term, such delays would have to be considered the norm rather than exceptions.

All these and other associated issues, when considered together, mean that while the demand for water in the arid and semi-arid countries would continue to increase steadily in the foreseeable future, arid countries are unlikely to have very many new sources of water which could be developed economically. In fact, for a large number of arid and semi-arid countries, international water bodies are the only major new source of water which could still be economically developed. Such water bodies have not been developed in the past because of the political complexities associated with their utilization. However, as water scarcities in individual countries become more and more serious, some countries may have no other alternative but to consider how best to use that resource, even though it could mean a 'beggar thy neighbour' attitude. This is why development and management of international water bodies would become an increasingly critical issue in the 1990s and beyond.

It is now evident that in the 1990s water will undoubtedly become a most critical resource for the future development and survival of the arid and semi-arid countries, so much so that all the indicators point to increasing tensions between neighbouring countries over the optimal use of international rivers, lakes and aquifers. Like the energy crisis of the 1970s, a serious water crisis now looms on the horizon. Unless every attempt is made to significantly improve the efficiency of existing water management processes, and all issues of utilization of various international water bodies are amicably and quickly resolved, the impending water crisis has the potential of becoming more pervasive and will adversely affect more lives than the energy crisis ever did even at its peak.

The main focus of the present paper is on the management of international water bodies; the issue of efficient water management has been discussed elsewhere (Biswas 1991; Thanh and Biswas 1990).

MAGNITUDE OF THE PROBLEM

The real magnitude of the problem of international water bodies is unknown at present. Even for international surface water bodies which are comparatively easy to identify, the real magnitude and extent of the problem is not known. The information base is significantly worse for international aquifers, since comparatively much less work has been carried out on such groundwater problems.

In the area of international river and lake basins, the first attempt to identify them was made by the Centre for Natural Resources, Energy and Transport (CNRET), now a defunct United Nations body. In its revised edition of the report *Integrated River Basin Development*, which was first published in 1958, it identified 166 international river basins on a world map.

In 1978, CNRET published a *Register of International Rivers*, which included information on lake basins as well. This publication identified 214 international river and lake basins, a number that was 29 per cent higher than the earlier estimate. This study defined a river basin as an 'area within which waters of natural origin (rain, groundwater flow, melting of snow and ice) feed a given river'. It considered only those river basins which were 'separate' (e.g. not tributary basins), and communicated 'directly with the final recipient of the water (oceans, closed inland seas or lakes)'.

The distribution of the 214 international river and lake basins by region is shown in Table 1.

Table 1. Distribution of international river and lake basins by region (CNRET 1978)

Region	Number
Africa	57
Asia	40
Europe	48
North and Central America	33
South America	36

According to the CNRET report, nearly 47 per cent of the area of the world (excluding Antarctica) falls within international basins, ranging from a high of nearly 60 per cent of the area in Africa and South America to a low of about 40 per cent in North and Central America. Detailed analysis of this report indicates that there are 44 countries where at least 80 per cent of the total area lies within international basins. Of these 44 countries, 20 are in Africa, 7 in Asia, 13 in Europe and 4 in Latin America.

The CNRET report, however, can only be considered to be a preliminary analysis of the problems. It certainly is not a definitive study, and suffers from many very serious methodological and factual short-comings. The real magnitude and extent of the problem of international rivers, according to this author, is significantly higher than this report indicates. Unfortunately, like many other environment and water development studies, the CNRET study has been quoted and requoted so many times that it is now accepted as a definitive analysis. This unquestionably is an erroneous conclusion for a variety of reasons, only the main ones of which will be discussed here.

First, the entire CNRET study was a desk study, which was based on maps available at the United Nations Map Library. As any experienced water planner knows, it is not an easy task to work on individual river basins only using maps, which often could be on a scale of 1:15,000,000, or even smaller. Reliable analysis and interpretation is a very difficult, if not impossible, task under the best of circumstances, especially for medium to small river basins. Generally it results in serious undercounting of such international basins.

Second, the study basically used the concept of topographical divides as basin boundaries. Unfortunately topographical divides do not necessarily indicate the direction of groundwater flow.

Third, the study used a planimeter to determine basin areas in different countries. This means that the reliability of the figures would depend directly on the reliability of the maps used, and also on the scales of these maps. Since it was only a desk study, and did not at all consider site investigations, all the errors generally went unchecked.

Fourth, there are many problems with the definition of what constitutes a basin. The methodological problems associated with an acceptable definition of an international river basin would be evident to anyone who has carefully followed the discussions of the India-Bangladesh Joint Rivers Commission. Equally, even if the so-called first-order basins are considered, as was the case for the CNRET study, it should be noted that many second- and third-order basins are larger than the first-order ones. Equally, certain smaller-order basins could be politically and in terms of water use more important than some first-order basins. Thus, for management of international water bodies, first-order basins are not necessarily more important than second- and lower-order ones.

Finally, the CNRET desk study was completed some 16 years ago, in 1976. During this period many new countries have been established in Eastern Europe. The break-up of the Soviet Union and other countries has now created new international river basins, which were earlier purely national in character.

Taking note of the above-mentioned points, it is evident that the number of international river basins in the world is significantly higher than the 214 identified by the U.N. study. A good example of this serious undercounting could be provided by the number of international rivers between India and Bangladesh. The U.N. study identified only one mega-basin, Ganges–Brahmaputra, which is shared not only by India and Bangladesh but also by China, Nepal and Bhutan. It should be noted that during one of the meetings of the India–Bangladesh Joint Rivers

Commission, Bangladesh identified more than 140 water systems that are common to both countries. Similarly, Nahid (1992) identifies 57 rivers that are common to these two countries.

It is evident that the earlier CNRET *Register of International Rivers* is now grossly out of date. We urgently need a more authoritative and up-to-date study which would provide a reliable picture of the extent of this problem globally. The CNRET study has had the unfortunate effect of reducing the perceived magnitude and extent of this major problem.

MANAGEMENT OF INTERNATIONAL RIVER BASINS: RECENT DEVELOPMENTS

It is submitted that management of international river basins has not received the attention it deserved during the past three decades. There have been some discussions at various international fora from time to time. However, these activities have been limited. Not only have they often lacked continuity, but also there has been very little coordination and integration of these limited activities undertaken by the various United Nations agencies and professional organizations. Thus, not surprisingly, very limited progress has been made during the past 30 years, either in terms of specifically solving the problems of individual river basins or in developing some acceptable rules which could be useful to countries attempting to resolve such difficult problems.

To the best of the author's knowledge, the first comprehensive study on the legal aspects of using the water of international rivers was carried out by Professor H. A. Smith of London. In his work *The Economic Uses of International Rivers*, published in 1931, he reviewed more than 100 treaties and studied several conflicts on the use of international rivers. He carefully refrained from making specific recommendations, which could be considered universal and thus used for resolution of conflicts between nations. He, however, emphasized the doctrine of riparian rights which entitled the lower riparian states to the natural flow of a river. He pointed out that some of the treaties analysed by him also considered the concept of equitable utilization.

In 1956, the International Law Association published the Dubrovnik rules for international rivers. Three years later, in 1959, Bolivia introduced a resolution in the General Assembly of the United Nations which requested the Secretary-General to prepare a report on laws related to international rivers. Thus, Resolution No. 1401 (XIV) of 21 November 1959, recommended that preliminary studies should be carried out on the problems associated with the development and use of international rivers in order to determine whether these could be codified. In response, the U.N. Secretary-General submitted two reports in 1963.

In 1966, the International Law Association (ILA), at its 52nd Conference held at Helsinki, adopted the so-called Helsinki rules for international watercourses. Four years later, in 1970, Finland introduced a resolution in the U.N. General Assembly on laws for international watercourses, which suggested that the Helsinki rules should be considered as a model.

The Sixth Committee of the U.N. discussed this proposal. While the Committee felt that the subject of international watercourse law was important, three reservations on the Helsinki rules surfaced. First, the rules were formulated by a professional organization which did not represent nation states. Second, some countries like Ethiopia argued that since nations had not participated in preparing the Helsinki Rules, adoption of these rules as a model could preclude new considerations on this complex issue. The third and probably the most important reservation was expressed about the fact that the Helsinki Rules were based on a drainage basin approach. Countries like Brazil, Belgium, China and France felt such an approach could be a potential threat to national sovereignty. They felt it was a radical departure from the traditional channel-based international law. In contrast, Finland and the Netherlands said the drainage basin framework was the most rational and scientific approach. Some countries considered that the problem of international river basins was so diverse that codification was not feasible.

The resolution to refer to the Helsinki Rules was lost (41 countries voted no, 25 voted yes and 32 abstained). It should be noted that this voting pattern was very unusual since it differed significantly from the then traditional pattern which was based on political alignments. After the deletion of reference to the Helsinki Rules, the resolution was passed with only one negative vote—that of Brazil. Thus, in Resolution No. 2669 (XIV) of 8 December 1970, the U.N. General Assembly noted that:

despite the great number of bilateral treaties and other regional regulations, as well as the Barcelona Convention of 1921 on the Regime of Navigable Waterways of Hydraulic Power affecting more than one state signed in Geneva in 1923, the use of international rivers and lakes is still based in part on general principles and rules of customary law.

The resolution also recommended that the International Law Commission should:

take up the study of the law of the non-navigational uses of international watercourses with a view to its progressive development and codification . . . .

Even though ILC included this subject in its programme of work in 1971, it was only in 1974 that a Sub-Committee was established to advise the ILC as to how best to proceed. The same year the Sub-Committee submitted a report which suggested that a questionnaire be circulated to the member governments on some key questions. The Commission accepted this proposal, and sent out a questionnaire in 1974 to all members of the General Assembly.

The questionnaire had 9 questions. One was on definition of the term 'international watercourse', two were on appropriateness of the drainage basin concept, five on what water uses and problems should be considered, and the last one was on the potential role of technical, scientific and economic experts.

The response to the questionnaire was not encouraging. By 1976, only 21 of the 147 U.N. members had bothered to reply. Four additional countries replied by 1978, one by 1979, four by 1980, and two by 1982. This meant that only about one-fifth of the member countries responded to a simple questionnaire in some 8 years!

Not surprisingly, the views of the countries on the appropriateness of the drainage basin concept—like the earlier discussion on the Helsinki Rules—were divided. Approximately half the countries supported the concept and the other half were either strongly negative or ambivalent. Argentina, Finland and the Netherlands supported the concept but Austria, Brazil and Spain opposed it strongly.

Because of countries' differing views, the Commission came to an agreement in 1976:

the question of determining the scope of the term 'international watercourses' need not be pursued at the outset of the work. Instead, attention should be devoted to beginning the formulation of general principles applicable to legal aspects of the uses of those watercourses.

The scope of the term was finally addressed by the Commission in 1991, under the guidance of the fourth Special Rapporteur, Prof. Stephen C. McCaffrey of the United States. The ILC adopted the draft articles on the Law of Non-Navigational Uses of Watercourses. These articles are given in the Appendix.

Between 1974 and 1991, the ILC had four Special Rapporteurs (three Americans and a Norwegian) to develop the draft laws. A fifth Special Rapporteur has recently been appointed. There are many reasons as to why it took some 21 years after the initial General Assembly request to have even the draft articles ready. First, the ILC depends completely on its Special Rapporteurs to prepare reports, which are then discussed. Since Special Rapporteurs have the full freedom to modify any approach and even withdraw previously adopted articles, changes in Rapporteurs could lead to a lengthening of the process. Second, membership of the Commission could change significantly every 5 years. For example, during the 1986 election, 14 of the 34 members elected were new, thus representing a 40 per cent turnover. Since new members may not be familiar with the subject or may have very different views compared to the countries they replaced, it could delay the process or even make the drafts internally inconsistent over a period of time. The terms of the members are not staggered, and thus continuity could be a serious constraint to speedy resolution of issues.

Progress at other U.N. fora: Discussions that were relevant to international water bodies were also carried out at other U.N. fora, especially those dealing with environmental issues. Thus, the United Nations Conference on the Human Environment, held at Stockholm in 1972, discussed certain aspects of natural resources that are shared by two or more countries. Principles 21 and 22 of the Declaration of that Conference dealt with this issue (United Nations, 1972).

According to Principle 21:

States have . . . the sovereign right to exploit their own resources pursuant to their own environmental policies, and the responsibility to ensure that activities within their jurisdiction or control do not cause damage to the environment of other States or of areas beyond the limits of national jurisdiction.

Similarly, Principle 22 stated:

States shall cooperate to develop further the international law regarding liability and compensation for the victims of pollution and other environmental damage caused by activities within the jurisdiction or control of such States to areas beyond their jurisdiction.

Five years later, in May 1977, the United Nations Water Conference held at Mar del Plata, Argentina, urged (Biswas 1978):

In relation to the use, management and development of shared water resources, national policies should take into consideration the right of each state sharing the resources to equitably utilize such resources as the means to promote the bonds of solidarity and cooperation.

A few months later, in September 1977, the United Nations

Conference on Desertification, held at Nairobi, Kenya, stated in Recommendation 26 under International Cooperation (United Nations 1978):

Experience has shown that processes of desertification at times transcend national boundaries, making efficient regional cooperation essential in the management of shared resources, with the objective of preventing ecological imbalance which can cause desertification.

In order to achieve judicious management and equitable sharing of resources on the basis of equality, sovereignty and territorial integrity, it is recommended that countries concerned should cooperate in the sound and judicious management of shared water resources as a means of combating desertification effectively.

The Desertification Conference reaffirmed the recommendation of the United Nations Water Conference that 'in the absence of bilaterial or multilateral agreements, Member States should continue to apply generally accepted principles of international law in the use, development and management of shared water resources' (United Nations 1978).

In spite of the above-mentioned declarations and resolutions, there has been very little progress on developing principles for the guidance of States in the management and harmonious use of shared natural resources. To some extent the lack of progress should not have been unexpected, especially if one reviewed what actually happened at both the Stockholm and Mar del Plata Conferences. The Stockholm recommendations on the destruction of tropical forests were insipid, primarily because certain countries, notably Brazil, strongly asserted that the use of forests, like other natural resources, was a matter of national decision-making only. Accordingly, deforestation recommendations finally approved were diluted and somewhat insipid: basically amounting to exhortations for further studies, surveys and data collection (Biswas and Biswas 1992).

The situation was somewhat different at the U.N. Water Conference, (1977), where international water bodies were implicitly considered to be a sensitive issue, and thus the discussions on this subject were very limited. The Secretary-General of the U.N. Water Conference, Yahia Abdel Mageed, noted five years later in a retrospective analysis:

. . . two other documents would have proved most useful in placing, more forcefully, before the Conference the questions of financial arrangements and shared water resources. It was felt that both these areas were not tackled satisfactorily at the Conference.

In addition to the afore-mentioned developments, the U.N. General Assembly adopted Resolution 3129 (XXVIII) on 13 December 1973, which stated:

Considers that it is necessary to ensure effective cooperation between countries through the establishment of adequate international standards for the conservation and harmonious exploitation of natural resources common to two or more states in the context of the normal relations existing between them;

Considers further that cooperation between countries sharing such natural resources and interested in their exploitation must be developed on the basis of a system of information and prior consultation within the framework of the normal relations existing between them.

The General Assembly then requested the Governing Council of the United Nations Environment Programme to 'report on measures adopted for their implementation'.

The Principle referred to above in the GA resolution was also endorsed at the Fourth Conference of Heads of State or Governments of Non-Aligned Countries at Algiers (5–9 September 1973) and later reconfirmed by article 3 of the Charter of Economic Rights and Duties of States as contained in the GA Resolution 3281 (XXIX).

In response to the GA resolution, UNEP established an Inter-governmental Working Group of Experts on Natural Resources shared by Two or More States, with the objective of preparing draft principles for the guidance of States. The discussions at the Group meetings were basically water-oriented. The Group worked from 1976 to 1978, and formulated 15 principles. These 'Draft Principles of Conduct' were formally approved by the Governing Council of UNEP on 24 May 1978, during its Sixth Session. The Governing Council authorized the Executive Director of UNEP to transmit the report to the General Assembly and invited 'the Assembly to adopt the draft principles'.

The issue was considered by the U.N. General Assembly in December 1978, but by then the situation had changed somewhat. The General Assembly resolution did not 'approve' the draft principles as the UNEP Governing Council had invited it to do; rather it merely 'took note' of the report and asked the U.N. Secretary-General 'to transmit the report to Governments for their study and comments' and then to report back to the General Assembly the following year. Thirty-four governments expressed their views, out of which 28 governments were in favour of adoption of the principles. The strongest criticisms came from Brazil ('give excuse for interference in environmental policies of sovereign States by outsiders'), Ethiopia ('vague, ambiguous, too general, incomplete and impractical'), and Japan ('doubts whether UNEP or the U.N. is the proper forum for dealing with this topic'). The U.N. Secretary General suggested that the principles be adopted, but the General Assembly in 1979 decided again only to 'take note' of the principles: it did not approve them.

In May 1982, the Governing Council of UNEP authorized its Executive Director to submit his report on cooperation in the field of environment concerning natural resources shared by two or more states to the General Assembly at its 37th session. It recommended to the General Assembly that the terms of the earlier Assembly resolution should be reiterated,

including its requests to all States to use the principles on the conservation and harmonious utilization of natural resources shared by two or more States as guidelines and recommendations in the formulation of bilateral and multilateral agreements regarding such resources.

In this context, it is interesting to note that the ILC adopted five articles on international waters in 1980, which included the concept that an international watercourse is a shared natural resource. Some upstream countries were not in favour of this concept, and also its implications were not clear. The concept of shared natural resources was eliminated shortly thereafter, and received no further consideration.

Other developments: In addition to the activities discussed earlier, other U.N. agencies and the World Bank have also carried out certain activities on international water bodies during the past four decades. Probably the most noteworthy and successful was the Indus River Treaty, which was signed on 19 September 1960 by India and Pakistan. This Treaty was clearly made possible by the foresight and leadership of Eugene Black, the then President of the World Bank.

Regrettably during the period 1960–80, the leadership shown by President Black was simply missing from all international organizations. Several reports were published, some meetings were convened, and certain missions were fielded by various U.N. agencies on international water bodies. Unfortunately their total impact was very limited, until the Executive Director of the United Nations Environment Programme, Dr Mostafa Kamal Tolba, initiated the Action Plan on the Zambezi River in the 1980s. Following the agreement on the Zambezi Action Plan, Dr Tolba expanded UNEP's interest to Lake Chad and the River Nile. Under the leadership of the UNEP, all the co-basin countries of the Nile, including Ethiopia, are now discussing how best to develop an environmentally-sound plan for the Nile Basin that would be acceptable to all the co-basin countries. The third meeting of the Nile Basin countries took place August 1992 in Nairobi. Unfortunately, the type of leadership shown by Black and Tolba could be considered as an exception rather than the rule during the past four decades.

ILC DRAFT

The ILC Draft on the law of the non-navigational uses of international watercourses contains 32 articles in six parts (see Appendix). Part I, entitled 'Introduction', contains 4 articles. Article 1 is on the scope of the draft. Article 2 defines certain terms used in the draft. Articles 3 and 4 are on watercourse agreements.

Part II outlines five 'General Principles'. They relate to equitable and reasonable utilization and participation (articles 5 and 6), obligation not to cause appreciable harm to other watercourses (article 7), general obligation for cooperation between watercourse states (article 8), regular exchange of data and information between states (article 9), and relationship between uses in the sense that 'no use of an international watercourse enjoys inherent priority over other uses' (article 10).

Part III is on 'Planned Measures', and contains 9 articles (articles 11–19). They primarily focus on the obligation of states to give prior notification and undertake the necessary consultation and negotiations with other concerned states on proposed new uses or changes in existing uses.

Part IV on 'Protection and Preservation' can be considered to be the environmental section of the draft. This part specifically is concerned with protection and preservation of ecosystems (article 20), prevention, reduction and control of pollution (article 21), introduction of alien or new species (article 22), and protection and preservation of the marine environment (article 23).

Part V is on harmful conditions (article 24) and emergency situations (article 25).

Part VI is entitled 'Miscellaneous Provisions' and contains 7 articles. They deal with joint management (article 26), regulation of the flow of waters (article 27), protection, maintenance and safe operation of installations, facilities and other works (article 28), international watercourses and installations in time of armed conflict (article 29), indirect procedures (article 30), data and information vital to national defence or security (article 31), and non-discrimination in terms of access to judicial and other procedures (article 32).

It should be noted that the Commission also had two additional parts on 'Implementation' and 'Fact-Finding and Settlement of Disputes'. But these two sections were not approved.

The Commission finally defined a watercourse as 'a system of surface and underground waters constituting by virtue of their physical relationship a unitary whole and flowing into a common terminus'.

The draft law is based on two fundamental principles. These are on equitable and reasonable utilization and obligation not to cause appreciable harm. According to Article 5:

Watercourse States shall in their respective territories utilize an international watercourse in an equitable and reasonable manner.

Similarly, Article 7 stipulates:

Watercourse States shall utilize an international watercourse in such a way as not to cause appreciable harm to other watercourse States.

There is no question that the draft rules are a step in the right direction. However, it is the first step of a long process and many issues need to be resolved if the draft articles are to be used for conflict resolution by countries sharing the various international water bodies.

Probably one of the most complex issues is the relation between the two main principles: equitable utilization and obligation not to cause harm. Goldberg (1991) points out:

whereas the injunction of not causing appreciable harm holds force as an imperative prohibition in absolute terms, the right to equitable sharing, although at times described as 'complimentary' is less readily conceived as self-standing inasmuch as the practical result in each case must be first determined by an agreement between the parties or an award of a competent tribunal. It is clear the right in question, i.e. to a reasonable and equitable sharing, involves a subjective judgement . . . .

Interestingly, the World Bank's policy on projects on international watercourses, which is outlined in its Operational Directive 7.50, firmly stipulates the 'no appreciable harm' principle but does not give similar emphasis to the concept of equitable sharing.

Stephen C. McCaffrey (1992), who was the last Special Rapporteur to guide the preparation of the draft rules, has raised four important questions on these rules:

1. Equitable utilization versus the obligation not to cause harm: which of these rules prevails in the event that they conflict?

2. What is the standard of responsibility for a breach of the draft articles—for example, article 7, which prohibits causing harm to other watercourse States?

3. Is the 'framework agreement' approach viable in the field of international watercourses?

4. Is the 'system' concept, as presently formulated in article 2, the soundest way of defining the physical scope of applicability of the draft articles (for example, should unrelated/confined groundwater have been included? Should the 'common-terminus' requirement be retained, and if so, should cases in which basins are connected by means of canals or otherwise somehow be taken into account)?

In addition to the four questions, the following five issues need further consideration.

1. The work of the ILC resulting in the preparation of the draft has contributed to a wealth of new information and ideas over the past two decades. However, rich though the work is on legal aspects, the process has suffered from the lack of good counsel on technical, economic and environmental issues. Clearly the problem of international water bodies can only be resolved through a multidisciplinary and holistic approach. Any uni-disciplinary attempt to resolve the problem is likely to produce suboptimal results on a long-term basis.

2. The draft has thus far not managed to integrate historical practice with emerging needs. Neither the ILC nor any other international institution has made a serious attempt to review the experiences of earlier agreements on international watercourses.

3. Limited attention has been paid to the work being carried out by other international and professional organizations in this area, except for the International Law Association.

4. There is no obligation under the rules to settle disputes according to any mechanism.

5. The environmental aspects suffer from the absence of an integrated ecosystems approach. In the present era of environmental awareness, this could prove to be a serious flaw.

The member states of the United Nations were expected to comment on the draft rules by January 1993. However, if past experiences are any indication, in all probability it will be a long time before a reasonable number of countries send their comments. In the mean time a new election has meant that many new countries are now members of the ILC for the next 5-year term. These countries may not be familiar with the discussions within the ILC, and further may not necessarily agree with all aspects of the draft. In addition, a new Special Rapporteur has been appointed. Thus, even though the General Assembly first asked in 1970 for a set of laws on non-navigational uses of international watercourses, only drafts have been formulated in 22 years. Many more years are likely to elapse before the drafts can be finalized.

GEOPOLITICS AND HYDROPOLITICS

In the area of management of international water bodies, geopolitical considerations and hydropolitical implications between the co-basin countries cannot be divorced from technical, legal, economic and environmental issues. When water becomes scarce and is considered to be a strategic national resource, hydropolitics needs to be reviewed for rational management of international water bodies.

In recent years, the strategic importance of water has often been compared with that of another liquid—oil. It is true that the geopolitics of oil is a critical issue. For example, if Kuwait was a major source of cabbages rather than oil, it is likely that Iraq's invasion would have been a very minor footnote in history. However, there is very little similarity between oil and water. For example, oil is only one major source of energy, but water has no substitute. Oil prices are very high when compared to water costs. Accordingly, it makes economic sense to transfer oil over very long distances, but not water. Also water consumption, especially for agricultural purposes, is significantly higher than oil consumption. In spite of these fundamental differences, it is water and not oil that has been attracting the attention of the world media in recent months, primarily due to geopolitical factors in the Middle East.

There are some social scientists and lawyers who have been arguing for some time for a water convention like the ones on ozone or climate change, since they feel water is no different from these environmental issues (Biswas 1992b). There are, however, some fundamental differences. Water is more controllable than ozone or climate. Ozone depletion and climate change will affect all nations, but the problems associated with individual international water bodies are very country-specific since only the countries concerned are parties to the dispute. For example, the Nile basin countries have very little, if any, interest in the management problems of Helmand or Karnafuli Rivers. Also, countries sharing an international water body can visualize the problems confronting them not only more tangibly but also directly in terms of perceived economic advantages. In contrast to ozone and climate change, it is much easier for the countries concerned to get excited over specific issues on the subject of water, and some 'sabre rattling' by the politicians could play well with the local populace. Equally, unlike the ozone issue, many co-basin countries consider international water bodies as the ultimate zero-sum game, and thus they often view each other as adversaries and not as partners. There are often historical grievances on such water bodies, and thus popular emotions can easily become inflamed within a very short period of time. Accordingly, irrespective of whether a water convention is desirable or not, it has to be admitted that such a convention is likely to have little similarity to ozone or climate change conventions.

Ever since the 1970 discussion in the U.N. General Assembly, many nations have expressed their reservations as to whether a framework convention on water would be useful or even possible. Having been involved with the negotiations on several international water bodies, a very likely scenario would be that not all the co-basin countries of a specific international water body are likely to sign a water convention, even though such a convention could be prepared for signature. A convention could add some moral pressure on recalcitrant countries, but on the basis of experience on the nuclear non-proliferation treaty, any such pressure, in the light of national self-interest, is likely to have limited impact. Accordingly, it is conceivable that legal codifications may not resolve *all* real-life problems, since the behaviour of nation states for the most part would depend upon their perceived political and economic self-interest.

CONCLUDING REMARKS

As the 21st century dawns, the issue of management of international water bodies will require more and more attention, both nationally and internationally. And yet, international organizations have for the most part tended to shy away from the resolution of specific problems because they are viewed as politically sensitive issues. To the extent they have become involved in such activities, the emphasis has been on data collection, exchange of information, sending of expert missions and convening seminars and conferences. The type of leadership shown by President Eugene Black of the World Bank in the 1950s and Mostafa Tolba of the UNEP in recent years stands in stark contrast to the 'softly, softly' approach of the international organizations. This attitude clearly has to change.

The root of the English word *rival* is from the Latin term *rivalis*, which originally meant using the same stream (*rivus*). But as the world becomes more interconnected, countries sharing the same river should no longer consider each other as rivals. It is not difficult to show that properly conceived management plans for international water bodies could result in win-win situations for all the parties concerned. Contary to popular belief, these are not zero-sum games. For the future welfare of mankind, the waters of international watercourses should be used optimally for the benefit of the people of all the concerned countries.

REFERENCES

Biswas, Asit K., 1978, *United Nations Water Conference: Summary and Main Documents*, Pergamon Press, Oxford, 217 pp.

——, 1991, 'Water for Sustainable Development in the 21st Century: A Global Perspective', Presidential Address to the International Water Resources Association, *Water International*, Vol. 16, No. 4, pp. 219–24.

——, 1992a, 'Freshwater Environment 1972–1992', *Water International*, Vol. 17, No. 2.

——, 1992b, 'Water for Third World Development: A Perspective from the South', *International Journal for Water Resources Development*, Vol. 8, No. 1, pp. 3–9.

Biswas, Margaret R., and Asit K. Biswas, 1992, 'Environment and Development in South Asia', *Contemporary South Asia*, Vol. 1, No. 2.

Centre for Natural Resources, Energy and Transport, 1978, 'Register of International Rivers', *Water Supply and Management*, Vol. 2, No. 1, pp. 1–58.

Goldberg, D., 1991, 'Legal Aspects of World Bank Policy on Projects on International Waterways', *International Journal for Water Resources Development*, Vol. 7, No. 4, pp. 225–29.

McCaffrey, S.C., 1992, 'Background and Overview of the International Law Commission's Study of the Non-Navigational Uses of International Watercourses', *Colorado Journal of International Environmental Policy and Law*, Vol. 3, No. 1, pp. 17–29.

Nahid Islam, 1992, 'Indo-Bangladesh Common Rivers: the Impact on Bangladesh', *Contemporary South Asia*, Vol. 1, No. 2.

Thanh, N.C., and Asit K. Biswas, 1990, *Environmentally-Sound Water Management*, Oxford University Press, New Delhi.

United Nations, 1973, *Report of the United Nations Conference on the Human Environment*, A/CONF. 48/14/Rev. 1, United Nations, New York.

United Nations, 1978, *United Nations Conference on Desertification: Round-up, Plan of Action and Resolutions*, United Nations, New York, p. 33.

Draft Report of the International Law Commission on the Work of its Forty-Third Session

The Law of the Non-Navigational Uses of International Watercourses[1]

PART I

Introduction

Article 1

Scope of the present articles

1. The present articles apply to uses of international watercourses and of their waters for purposes other than navigation and to measures of conservation related to the uses of those watercourses and their waters.

2. The use of international watercourses for navigation is not within the scope of the present articles except in so far as other uses affect navigation or are affected by navigation.

Article 2

Use of terms

For the purposes of the present articles:

(a) 'international watercourse' means a watercourse, parts of which are situated in different States;

(b) 'watercourse' means a system of surface and underground waters constituting by virtue of their physical relationship a unitary whole and flowing into a common terminus;

(c) 'watercourse State' means a State in whose territory part of an international watercourse is situated.

---

[1] *Draft Articles on the Law of the Non-Navigational Uses of International Watercourses, Draft Report of the International Law Commission*, U.N. GAOR, 43rd Sess., at 1, U.N. Doc. A/CN/.4/L.463/Add.4 (1991).

Article 3

Watercourse agreements

1. Watercourse States may enter into one or more agreements, hereinafter referred to as 'watercourse agreements', which apply and adjust the provisions of the present articles to the characteristics and uses of a particular international watercourse or part thereof.

2. Where a watercourse agreement is concluded between two or more watercourse States, it shall define the waters to which it applies. Such an agreement may be entered into with respect to an entire international watercourse or with respect to any part thereof or a particular project, programme or use, provided that an agreement does not adversely affect, to an appreciable extent, the use by one or more other watercourse States of the waters of the watercourse.

3. Where a watercourse State considers that adjustment or application of the provisions of the present articles is required because of the characteristics and uses of a particular international watercourse, watercourse States shall consult with a view to negotiating in good faith for the purpose of concluding a watercourse agreement or agreements.

Article 4

Parties of watercourse agreements

1. Every watercourse State is entitled to participate in the negotiation of and to become a party to any watercourse agreement that applies to the entire international watercourse, as well as to participate in any relevant consultations.

2. A watercourse State whose use of an international watercourse may be affected to an appreciable extent by the implementation of a proposed watercourse agreement that applies only to a part of the watercourse or to a particular project, programme or use is entitled to participate in consultations, and in the negotiation of, such an agreement, to the extent that its use is thereby affected, and to become a party thereto.

PART II

General Principles

Article 5

Equitable and reasonable utilization and participation

1. Watercourse States shall in their respective territories utilize an international watercourse in an equitable and reasonable manner. In

particular, an international watercourse shall be used and developed by watercourse States with a view to attaining optimal utilization thereof and benefits therefrom consistent with adequate protection of the watercourse.

2. Watercourse States shall participate in the use, development and protection of an international watercourse in an equitable and reasonable manner. Such participation includes both the right to utilize the watercourse and the duty to cooperate in the protection and development thereof, as provided in the present articles.

Article 6

Factors relevant to equitable and reasonable utilization

1. Utilization of an international watercourse in an equitable and reasonable manner within the meaning of article 5 requires taking into account all relevant factors and circumstances, including:

(a) geographic, hydrographic, hydrological, climatic, ecological and other factors of a natural character;

(b) the social and economic needs of the watercourse States concerned;

(c) the effects of the use or uses of the watercourse in one watercourse State on other watercourse States;

(d) existing and potential uses of the watercourse;

(e) conservation, protection, development and economy of use of the water resources of the watercourse and the costs of measures taken to that effect;

(f) the availability of alternatives, of corresponding value, to a particular planned or existing use.

2. In the application of article 5 or paragraph 1 of this article, watercourse States concerned shall, when the need arises, enter into consultations in a spirit of cooperation.

Article 7

Obligation not to cause appreciable harm

Watercourse States shall utilize an international watercourse in such a way as not to cause appreciable harm to other watercourse States.

Article 8

General obligation to cooperate

Watercourse States shall cooperate on the basis of sovereign equality, territorial integrity and mutual benefit in order to attain optimal utilization and adequate protection of an international watercourse.

Article 9

Regular exchange of data and information

1. Pursuant to article 8, watercourse States shall on a regular basis exchange reasonably available data and information on the condition of the watercourse, in particular that of a hydrological, meteorological, hydrogeological and ecological nature, as well as related forecasts.

2. If a watercourse State is requested by another watercourse State to provide data or information that is not reasonably available, it shall employ its best efforts to comply with the request but may condition its compliance upon payment by the requesting State of the reasonable costs of collecting and, where appropriate, processing such data or information.

3. Watercourse States shall employ their best efforts to collect and, where appropriate, to process data and information in a manner which facilitates its utilization by the other watercourse States to which it is communicated.

Article 10

Relationship between uses

1. In the absence of agreement or custom to the contrary, no use of an international watercourse enjoys inherent priority over other uses.

2. In the event of a conflict between uses of an international watercourse, it shall be resolved with reference to the principles and factors set out in articles 5 to 7, with special regard being given to the requirements of vital human needs.

PART III

Planned Measures

Article 11

Information concerning planned measures

Watercourse States shall exchange information and consult each other on the possible effects of planned measures on the condition of an international watercourse.

Article 12

Notification concerning planned measures with possible adverse effects

Before a watercourse State implements or permits the implementation of planned measures which may have an appreciable adverse effect upon other watercourse States, it shall provide those States with timely notification thereof. Such notification shall be accompanied by available technical data and information in order to enable the notified States to evaluate the possible effects of the planned measures.

Article 13

Period for reply to notification

Unless otherwise agreed, a watercourse State providing a notification under article 12 shall allow the notified States a period of six months within which to study and evaluate the possible effects of the planned measures and to communicate their findings to it.

Article 14

Obligations of the notifying State during the period for reply

During the period referred to in article 13, the notifying State shall cooperate with the notified States by providing them, on request, with any additional data and information that is available and necessary for an accurate evaluation, and shall not implement or permit the implementation of the planned measures without the consent of the notified States.

Article 15

Reply, to notification

1. The notified States shall communicate their findings to the notifying State as early as possible.

2. If a notified State finds that implementation of the planned measures would be consistent with the provisions of articles 5 or 7, it shall communicate this finding to the notifying State within the period referred to in article 13, together with a documented explanation setting forth the reasons for the finding.

Article 16

Absence of reply to notification

If, within the period referred to in article 13, the notifying State receives no communication under paragraph 2 of article 15, it may, subject to its obligations under articles 5 and 7, proceed with the implementation of the planned measures, in accordance with the notification and any other data and information provided to the notified States.

Article 17

Consultations and negotiations concerning planned measures

1. If a communication is made under paragraph 2 of article 15, the notifying State and the State making the communication shall enter into consultations and negotiations with a view to arriving at an equitable resolution of the situation.

2. The consultations and negotiations shall be conducted on the basis that each State must in good faith pay reasonable regard to the rights and legitimate interests of the other State.

3. During the course of the consultations and negotiations, the notifying State shall, if so requested by the notified State at the time it makes the communication, refrain from implementing or permitting the implementation of the planned measures for a period not exceeding six months.

Article 18

Procedures in the absence of notification

1. If a watercourse State has serious reason to believe that another watercourse State is planning measures that may have an appreciable adverse effect upon it, the former State may request the latter to apply the provisions of article 12. The request shall be accompanied by a documented explanation setting forth the reasons for such belief.

2. In the event that the State planning the measures nevertheless finds that it is not under an obligation to provide a notification under article 12, it shall so inform the other State, providing a documented explanation setting forth the reasons for such finding. If this finding does not satisfy the other State, the two States shall, at the request of that other State, promptly enter into consultations and negotiations in the manner indicated in paragraphs 1 and 2 of article 17.

3. During the course of the consultations and negotiations, the State planning the measures shall, if so requested by the other State at the time it requests the initiation of consultations and negotiations, refrain from implementing or permitting the implementation of those measures for a period not exceeding six months.

Article 19

Urgent implementation of planned measures

1. In the event that the implementation of planned measures is of the utmost urgency in order to protect public health, public safety or other

equally important interests, the State planning the measures may, subject to articles 5 and 7, immediately proceed to implementation, notwithstanding the provisions of article 14 and paragraph 3 of article 17.

2. In such cases, a formal declaration of the urgency of the measures shall be communicated to the other watercourse States referred to in article 12 together with the relevant data and information.

3. The State planning the measures shall, at the request of any of the States referred to in paragraph 2, promptly enter into consultations and negotiations with it in the manner indicated in paragraphs 1 and 2 of article 17.

PART IV

Protection and Preservation

Article 20

Protection and preservation of ecosystems

Watercourse States shall, individually or jointly, protect and preserve the ecosystems of international watercourses.

Article 21

Prevention, reduction and control of pollution

1. For the purposes of this article, 'pollution of an international watercourse' means any detrimental alteration in the composition or quality of the waters of an international watercourse which results directly or indirectly from human conduct.

2. Watercourse States shall, individually or jointly, prevent, reduce and control pollution of an international watercourse that may cause appreciable harm to other watercourse States or to their environment, including harm to human health or safety, to the use of the waters for any beneficial purpose or to the living resources of the watercourse. Watercourse States shall take steps to harmonize their policies in this connection.

3. Watercourse States shall, at the request of any of them, consult with a view to establishing lists of substances, the introduction of which into the waters of an international watercourse is to be prohibited, limited, investigated or monitored.

Article 22

Introduction of alien or new species

Watercourse States shall take all measures necessary to prevent the introduction of species, alien or new, into an international watercourse which may have effects detrimental to the ecosystem of the watercourse resulting in appreciable harm to other watercourse States.

Article 23

Protection and preservation of the marine environment

Watercourse States shall, individually or jointly, take all measures with respect to an international watercourse that are necessary to protect and preserve the marine environment, including estuaries, taking into account generally accepted international rules and standards.

PART V

Harmful Conditions and Emergency Situations

Article 24

Prevention and mitigation of harmful conditions

Watercourse States shall, individually or jointly, take all appropriate measures to prevent or mitigate conditions that may be harmful to other watercourse States, whether resulting from natural causes or human conduct, such as flood or ice conditions, water-borne diseases, siltation, erosion, salt-water intrusion, drought or desertification.

Article 25

Emergency situations

1. For the purposes of this article, 'emergency' means a situation that causes, or poses an imminent threat of causing, serious harm to watercourse States or other States and that results suddenly from natural causes, such as floods, the breaking up of ice, landslides or earthquakes, or from human conduct as for example in the case of industrial accidents.

2. A watercourse State shall, without delay and by the most expeditious means available, notify other potentially affected States and competent international organizations of any emergency originating within its territory.

3. A watercourse State within whose territory an emergency originates shall, in cooperation with potentially affected States and, where appropriate, competent international organizations, immediately take all practicable measures necessitated by the circumstances to prevent, mitigate and eliminate harmful effects of the emergency.

4. When necessary, watercourse States shall jointly develop contingency plans for responding to emergencies, in cooperation, where appropriate, with other potentially affected States and competent international organizations.

PART VI

Miscellaneous Provisions

Article 26

Management

1. Watercourse States shall, at the request of any of them, enter into consultations concerning the management of an international watercourse, which may include the establishment of a joint management mechanism.

2. For the purposes of this article, 'management' refers, in particular, to:

(a) planning the sustainable development of an international watercourse and providing for the implementation of any plans adopted; and

(b) otherwise promoting rational and optimal utilization, protection and control of the watercourse.

Article 27

Regulation

1. Watercourse States shall cooperate where appropriate to respond to needs or opportunities for regulation of the flow of the waters of an international watercourse.

2. Unless they have otherwise agreed, watercourse States shall participate on an equitable basis in the construction and maintenance or defrayal of the costs of such regulation works as they may have agreed to undertake.

3. For the purposes of this article, 'regulation' means the use of hydraulic works or any other continuing measure to alter, vary or otherwise control the flow of the waters of an international watercourse.

Article 28

Installations

1. Watercourse States shall, within their respective territories, employ their best efforts to maintain and protect installations, facilities and other works related to an international watercourse.

2. Watercourse States shall, at the request of any of them which has serious reason to believe that it may suffer appreciable adverse effects, enter into consultations with regard to:

(a) the safe operation or maintenance of installations, facilities or other works related to an international watercourse; or

(b) the protection of installations, facilities or other works from wilful or negligent acts or the forces of nature.

Article 29

International watercourses and installations in time of armed conflict

International watercourses and related installations, facilities and other works shall enjoy the protection accorded by the principles and rules of international law applicable in international and internal armed conflict and shall not be used in violation of those principles and rules.

Article 30

Indirect procedures

In cases where there are serious obstacles to direct contacts between watercourse States, the States concerned shall fulfil their obligations of cooperation provided for in the present articles, including exchange of data and information, notification, communication, consultations and negotiations, through any indirect procedures accepted by them.

Article 31

Data and information vital to national defence or security

Nothing in the present articles obliges a watercourse State to provide data or information vital to its national defence or security. Nevertheless, that State shall cooperate in good faith with the other watercourse States with a view to providing as much information as possible under the circumstances.

Article 32

Non-discrimination

Watercourse States shall not discriminate on the basis of nationality or residence in granting access to judicial and other procedures, in accordance with their legal systems, to any natural or juridical person who has suffered appreciable harm as a result of an activity related to an international watercourse or is exposed to a threat thereof.

Index